Geomagnetic Observatory and Survey Practice

Edited by

W. F. STUART
British Geological Survey, Edinburgh

Reprinted from

Geophysical Surveys, Vol. 6, Nos. 3/4

D. Reidel Publishing Company

Dordrecht / Boston

ISBN-13: 978-94-010-8833-6

Softcover reprint of the hardcover 1st edition 1984

ISBN-13:978-94-010-8833-6 e-ISBN-13:978-94-009-5283-6
DOI: 10.1007/978-94-009-5283-6

GEOMAGNETIC OBSERVATORY AND SURVEY PRACTICE

Edited by

W. F. STUART

W. F. STUART / Introduction 217

E. KRING LAURIDSEN / Absolute Measurement of *D* by Means of a Proton Magnetometer 223

J. BITTERLY, J. M. CANTIN, R. SCHLICH, J. FOLQUES, and D. GILBERT / Portable Magnetometer Theodolite with Fluxgate Sensor for Earth's Magnetic Field Component Measurements 233

H.-J. LINTHE and D. LENNERS / Acquisition and Primary Processing of Geomagnetic Pulsations in Real Time Operation by Microcomputer 241

B. J. FRASER and P. W. McNABB / The Recording of Pc1 Geomagnetic Pulsations Using a Microcomputer Preprocessing System 249

J. PODSKLAN / Magnetic Mapping of Slovakia for the Epoch 1980.5 261

V. AUSTER / Geomagnetic Absolute Measurements with a Nuclear Resonance Theodolite at the Adolf Schmidt Observatory in Niemegk 269

S. UTASHIRO, S. OSHIMA, and T. KANEKO / Aeromagnetic Surveys in the Seas Around Japan in 1980 271

G. CLERC, J-P. DÉCRIAUD, G. DOYEN, M. HALBWACHS, M. HENROTTE, J. RÉMY, and X-C. ZHANG / An Automatic Audio-Magnetotelluric Equipment, Controlled by Microprocessor, for the Telesurveillance of the Volcano Momotombo (Nicaragua) 291

H. LÜHR, S. THÜREY, and N. KLÖCKER / The Eiscat-Magnetometer Cross 305

L. HEGYMEGI and L. DRIMUSZ / An Intelligent Digital Magnetic Recording System (Dimars) 317

GÜNTHER SCHULZ and MARTIN BEBLO / On the Reduction of Time Variations for Geomagnetic Repeat Station Measurements 323

J. PODSKLAN and I. KOVAC / One Year Experience with the CMVS 2 Magnetic Variation Station 335

A. MELONI, F. MOLINA, P. PALANGIO, Q. TACCETTI, and ANNA DE SANTIS / Automatic Digital Recording of Geomagnetic Elements by Means of a Proton Precession Magnetometer 339

V. AUSTER and K. LENGNING / Comparison of Standard and Digital Observation at the Geomagnetic Adolf Schmidt Observatory Niemegk 351

M. KUWASHIMA and Y. SANO / Improved Kakioka Automatic Standard Magnetometer (KASMMER) 357

J. JANKOWSKI, J. MARIANIUK, A. RUTA, C. SUCKSDORFF, and M. KIVINEN / Long-Term Stability of a Torque-Balance Variometer with Photoelectric Converters in Observatory Practice 367

TORSTEN BERGMARK / Experience of Geomagnetic Field Recording with a Fluxgate Magnetometer Having a Bridge Sensor 381

A. J. FORBES and J. C. RIDDICK / The Digital Recording System Operated at the U.K. Magnetic Observatories 393

D. A. SIMMONS and J. R. ROUSE / Geomagnetic Measurements Made on the Moving Ice Shelf at Halley, Antarctica 407

J. A. JOSELYN / Proposed Major Format Change to Geomagnetic Activity Reports and Forecasts Produced by the SESC, Boulder, Colorado, U.S.A. 419

ASGER LUNDBAK / About Digital Alternative to the *Kp*-Index 425

E. R. NIBLETT, E. I. LOOMER, R. L. COLES, and G. JANSEN VAN BEEK / Derivation of *K*-Indices Using Magnetograms Constructued from Digital Data 431

J. C. RIDDICK and W. F. STUART / The Generation of *K* Indices from Digitally Recorded Magnetic Data 439

Announcement 457

INTRODUCTION

This issue is a collection of the papers read at the 'Workshop on Geomagnetic Observatory and Survey Practice' held during the XIVth General Assembly of IUGG (the International Union of Geology and Geophysics) in Hamburg, August 1983, sponsored by Division V of the International Association of Geomagnetism and Aeronomy (IAGA). The papers represent a snapshot taken at a very important time in the history of Geomagnetism and of the sciences which depend on measurements of one kind or another of the Earth's magnetic field. Research science now demands a much greater amount of information to be prepared and immediately made available to the scientific user. Experimental measurements are now required to be reduced, selected and made ready as information which can be recorded as data on magnetic tape in the form required for direct incorporation into the analytical programmes which individual researchers run on digital computers. Computing has reduced the lead time between when observations are made and when they are required by researchers. Many scientific programmes, particularly those related to Solar-terrestrial geophysics, need data to be analysed as near as possible to the time it is recorded. In Geomagnetism these pressures apply to field variations where satellite based geophysical experiments require high resolution of the fine structure of external disturbance fields, and also to field mapping on a global and local scale where the demand for increased accuracy calls for better absolute observations and more frequent surveys. Applications include predicting the state of the magnetosphere from the magnetic weather pattern, and determining geological structures in the crust from the effects of electromagnetic induction taking place in the Earth's Core from secular variation patterns around the globe.

Geomagnetism's history began with the evolution of the magnetic compass which was in use in China perhaps as early as 600 BC (arguably the first scientific instrument). The properties of the compass were the subject of a great deal of theorising by European philosophers between the 12th and 16th centuries AD when observations began to show that its directional properties were due to the planet Earth. In 1600 William Gilbert summarised work of his own and much that had preceded him in a paper called 'De Magnete' which identified the Declination and Inclination of a freely suspended magnet, showed how they varied at different locations on the Earth and concluded that the Earth itself is no more than a large magnet. Declination results from the fact that the magnetic axis of the Earth does not coincide with the geographical axis; thus a compass shows a false bearing which varies with longitude. To navigators the correction which is needed and its variation around the world is of prime significance. To this day sailors and most map makers refer to magnetic Declination as the Variation of the compass.

The first magnetic chart of the world (showing Declination) was published in 1702 following Halley's survey of the Atlantic a few years earlier. It laid the basis of an increasing demand by navigators for global and local magnetic field information because it had been realised that, among other things, the position of the magnetic poles

Geophysical Surveys **6** (1984) 217.

changes slowly with time thus rendering charts obsolete almost as they are produced.

The earliest magnetic observations were made by carefully suspended magnets. In the early part of the 19th century Gauss evolved the field theory of magnetism and devised the methods for measuring the geomagnetic field by suspended magnets carefully calibrated against electric currents. He developed the mathematics of modelling the Earth's magnetic field and initiated a programme of measurements and recordings globally which has become a network of permanent magnetic observatories at fixed sites, supplemented by surveys where the field is measured from time to time over large areas. The methods devised by Gauss continue to be used today and observatory instruments are not much more than improvements in the basic designs of one or two centuries ago.

When the proton precession magnetometer was developed 30 years ago it was hoped that it and other atomic and nuclear magnetometers would lead the way to a new generation of instruments capable of truly absolute measurements of magnetic field and of being adapted to completely automatic methods of observing. Exhaustive investigation has been negative, although the proton resonance has been adopted as the standard reference for the geomagnetic field and the Proton magnetometer itself is recognised as the only device for measuring absolute fields. However technology is now available to allow any suitable sensor to be the basis of a digitally recording variometer and if techniques can be developed to make good enough absolute measurements fully automatic monitoring of the geomagnetic field is conceivable at observations and in field surveys. Fluxgates sensors and photo-electrically recording magnet suspensions have emerged as the best sensors.

The second half of the Workshop contains papers which summarise the instrumental and technical development taking place in many parts of the world as a step in converting magnetic observatory practice from classical variometers and handscaling methods to digital recording and computer data processing. The work reported here will be the basis for a much improved service to the geomagnetic community in the next few years. Trials of automatic observatories were undertaken in the late 1960's by Britain, Canada and the United States but only Canada, under the direction of Paul Serson, adopt digital recording as its principal at an extensive network of observatories in the early 1970's. Since then most countries have watched with interest the outcome of those early trials. Economic pressure has now tipped the balance and most advanced countries are preparing to accept the possibility of lower overall standard of observation in the interest of less labour-intensive digital recording and data processing.

By far the greater volume of data (and research papers) required by the scientific community is that relating to field variations without reference to the absolute magnitude of the field itself. Digital recording will not reduce the quality of this information and will improve the service to research immensely. Absolute measurements are complex things because the field varies constantly; the slow secular variation of the Core, rythmical external variations daily and annually and less predictable disturbances. Thus only by making a series of absolute measurements and eliminating most of the variation effects can a true absolute value be achieved. Even then, since the process takes time, there will remain a systematic error due to the secular variation. It is in this process of arriving at the absolute field value that some loss of observing standard may appear in purely automatic observatories.

The papers in the opening section deal with the instrumental and technical problems which have to be overcome. The references to baseline stability, standard deviation of mean values and level of drop outs refer as much to the authors' preoccupation with meeting accepted classical standards as they do to simple instrument specifications. The standard of operation of magnetic observations is an exacting one in the sensitivity and accuracy with which the field is recorded and the high level of continuity of record that is produced. The loss of standard which unattended operation introduces has not yet been assessed. That can only be done when several years mean annual values are available. The prognosis is good within the range of standards achieved overall by the worlds' observatories. Continuity presents a greater problem. The continuity of an analogue photographic record is very difficult to achieve in modern digital equipment where data may be transmitted over large distances. Perhaps of more concern is that there should not be too much discontinuity in the standard of recording and the interpretation of the records.

Digital recording, by its nature, is low pass filtered and the effects of discrete sampling rates on the data must be carefully examined. In a later section papers examine the K index to see if it can be produced from digital data in the same form as originally intended. Created in 1934 by Bartels in 1939 the significance of K is that it estimates non-systematic variations of the external field. To achieve this an observer who has a thorough knowledge of the Daily Variation and its day to day and seasonal changes must estimate its effect in the magnetic record. Digital recording implies that familiarity with the daily record will be hard to achieve and in fully automatic data collection and processing the daily record will hardly ever be inspected. It is unlikely that an automatic process can estimate Daily Variation sufficiently well to duplicate the K index and it becomes necessary to consider replacing K with an equivalent index which is acceptable for the science to which it is to be applied. The papers here show that digitally produced indices are very close to the ideal. For the vast majority of science they are perfectly adequate and to many the question becomes whether a better index can be provided for the scientific user now that the power of digital computation can be applied to the data.

The importance of this collection of papers is that it represents a summary of the situation which will be the basis of the transition from 'classical' to digital procedure at geomagnetic observatories. The scientific research community must now respond by saying how it can best use the new style data and feeding back research discoveries so that the most effective observing and data processing procedures can be adopted by the observatories.

It is apparent that full automation will be a fairly rapid consequence of digital recording at many observatories. Economic and technological factors make this inevitable. Automation will thus strengthen the worldwide network because observatories may be operated where it was never possible before (remote oceanic islands perhaps) or where local manpower limitations led to irregular observations or erratic operation. Digital recording will improve vastly the transmission of data to World Data Centres for dissemination abroad.

In the preparation of the mathematical model for the World Magnetic Chart data from less than half the total number of observatories are up to date (i.e. about one year behind the date of compilation). Of the remaining, one quarter of the worlds observatories are up to five or six years behind in the production of these data and the rest are even worse. The adoption of digital recording eliminates any real excuse for such poor reproduction of what is one fundamental measurement of the planet Earth.

The next step in the process of evolution of magnetic observatories will be at Prague in August 1984 when IAGA holds its 5th Scientific Assembly. Another Observatory Workshop will discuss the progress made since these papers were written and there will be a Symposium on the implications to research of digital recording. While that is going on meetings will be taking place planning magnetic survey satellites, follow up missions to MAGSAT which orbitted the Earth for almost six months in 1980 mapping the geomagnetic field in great detail. MAGSAT produced a high resolution snapshot of the field, with many complicated external field errors added and without a really reliable measure of secular variation. When a magnetic survey satellite can be flown at regular intervals (once or twice per year) for many years in succession the magnetic observatory system will have completed its technological revolution. Ground observatories will be retained only as a base for local survey operations, activity indices and short term variations experiments. The experience gained now in handling the data, locally and on a global basis will lay the groundwork for the regular use of satellite surveys for the geomagnetic mapping of the future.

W. F. STUART

Geomagnetism Research Group
British Geological Survey
West Mains Road
Edinburgh, Scotland, EH9 3LA

The first few papers deal with basic instrument developments. They illustrate the fact that the work continues to be wideranging from techniques of measuring absolute values of the magnetic field under observatory conditions to microprocessor control systems recording rapid variations and selecting specific events for analysis. Progress in absolute observations is predicatably slow and painstaking by the nature of the problem. Technology alone controls the rate of progress made in variations recording and the use of dedicated microprocessors to compress and automate data collection and analysis will lead to the recording of characteristics of rapid geomagnetic variations on a routine basis which would otherwise be impossible. Such characteristics may provide the magnetic indices of the future and contribute directly to the ability to predict magnetospheric and ionsopheric behaviour from magnetic activity.

ABSOLUTE MEASUREMENT OF *D* BY MEANS OF A PROTON MAGNETOMETER

E. KRING LAURIDSEN
Danish Meteorological Institute, DK-2100 Copenhagen

Abstract. In a large coil with vertical axis the current is adjusted so that $Z = 0$. A smaller coil of a new type with four sets of cylindrical turns is placed with its axis horizontal. This *D*-coil is provided with a telescope pointing at a mark nearly in the magnetic meridian. A proton magnetometer sensor is placed in the common centre of the coils. Two series of readings are taken with the *D*-coil in the erect and inverted position respectively. Variations of *D*, *H*, *F*, and *D*-coil current are recorded. A simple formula gives the mean value of *D*.

1. Introduction

Previously, I have described how the intrinsic geomagnetic elements *F*, *H*, and *Z* are measured at the Danish geomagnetic observatory at Brorfelde with a proton magnetometer (Kring Lauridsen, 1980). The principle is the compensation method, but we use stationary coils. The possible misalignment of the coil axis is measured directly by means of a variometer with liquid damping, and a corresponding correction to the measured value is applied.

There are some advantages of this method compared to the usual method with turning coils:

(1) no problems with a mechanical axis of the base and accuracy of levels;
(2) large square coils can be manufactured easily from commercial profiles;
(3) the main pier is left free for measurements with other instruments;
(4) likewise the coils are available for other purposes.

The present paper extends the method allowing absolute measurement of declination *D* by means of an extra coil. Whereas the coils for the compensation method are large coils (square coils with edge length about 2 m) the extra *D*-coil is a rather small one (a

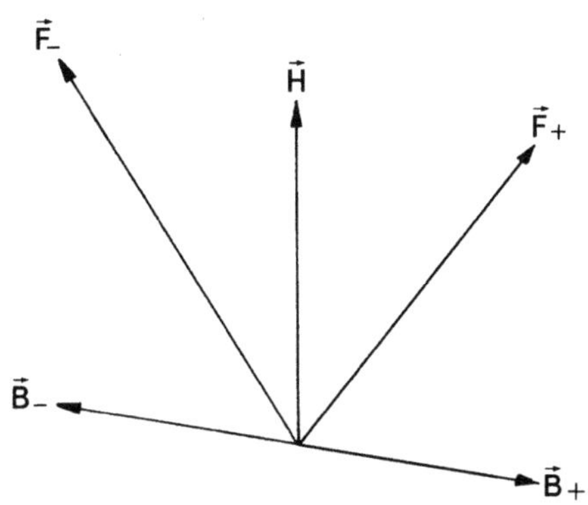

Fig. 1. Resulting total field from ambient horizontal field *H* and coil field *B* wit current normal and reversed.

Geophysical Surveys **6** (1984) 223-232. 0046-5763/84/0063-0223$01.50.

new cylindrical type proposed by Robinson (1983) of diameter about 1/3 m) thus allowing inversion of the coil.

The principle is to create an extra field nearly perpendicular to the magnetic meridian. If this extra horizontal coil field B is added to the ambient horizontal field H (Z is assumed to be annulled) the resultant is $F_+ = |\mathbf{H} + \mathbf{B}|$ which is measured by the proton magnetometer, cf. Figure 1. If the current in the coil is reversed the new resultant $F_- = |\mathbf{H} - \mathbf{B}|$ will differ numerically from F_+ unless $\mathbf{B} \cdot \mathbf{H} = 0$. In this case it can be concluded that the magnetic meridian is perpendicular to the coil axis.

Now, the magnetic axis of a coil does not manifest itself. In order to have a reference of the direction of the coil it may be provided with a telescope. Thereby an unknown angle of colimation is introduced. However, this angle can be eliminated by inversion of the coil, as is also done in measurements of D by means of a declinometer, for instance QD (Kring Lauridsen, 1981), cf. Figure 2.

It is not necessary to find the exact position where the coil axis is perpendicular to the meridian just as the position of no deflection of the galvanometer is hunted by an earth-inductor. When the variations of D, H, F, and B are recorded it is possible to calculate the exact position of the meridian by means of a simple formula. The theory of the procedure is explained in the following section. The telescope which is fixed to the coil perpendicular to the axis is pointed at a reference mark approximately in the meridian both in the erect and the inverted position of the coil.

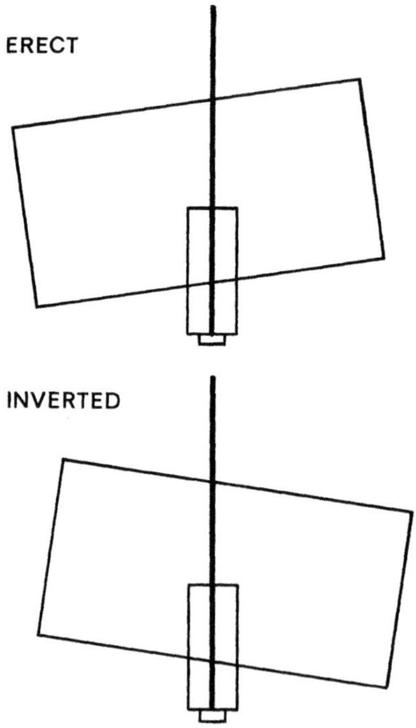

Fig. 2. The collimation angle between the optical axis of the telescope and the coil axis is eliminated by inversion of the coil, and realignment of the telescope axis.

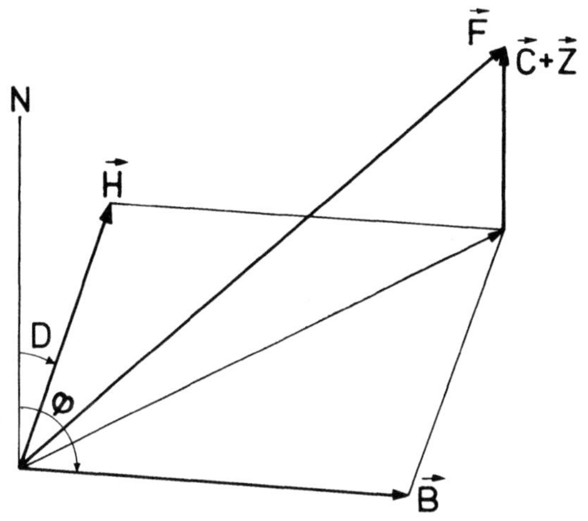

Fig. 3. The *D*-coil creates an artificial horizontal field *B* with azimuth φ and a vertical field *C*.

2. Fundamental Theory

The coil creates a horizontal field *B* with azimuth φ (cf. Figure 3) and a possible vertical field *C*. The proton magnetometer sensor is placed in the centre of this coil and monitors the total field *F*. We assume that $\varphi_e - D_e \approx \pi/2$ in the erect position, and that $\varphi_i - D_i \approx -\pi/2$ in the inverted position. The two positions are indicated by indices *e* and *i* respectively (*e* for erect and east).

The total field *F* resulting from vector addition of the ambient and the artificial field is given by

$$F_e^2 = H_e^2 + B_e^2 + 2H_e B_e \cos(\varphi_e - D_e) + (Z_e + C_e)^2. \tag{1}$$

The inverted position results from the erect position by lifting the coil and turning its axis in the opposite direction. In this position

$$F_i^2 = H_i^2 + B_i^2 + 2H_i B_i \cos(\varphi_i - D_i) + (Z_i + C_i)^2. \tag{2}$$

On the assumption that the coil axis is pointing approximately magnetic east in the erect position and magnetic west in the inverted position we have

$$\varphi_e - D_e \approx \pi/2; \qquad \varphi_i - D_i \approx -\pi/2 \tag{3}$$

implying that

$$\cos(\varphi_e - D_e) \approx \pi/2 - (\varphi_e - D_e); \qquad \cos(\varphi_i - D_i) \approx \pi/2 + (\varphi_i - D_i). \tag{4}$$

These factors are extremely small and therefore we can neglect possible variations in the values of the factors *H* and *B* in the corresponding terms. Thus (1) and (2) can be replaced with good approximation by

$$F_e^2 = H_e^2 + B_e^2 + 2HB\,(\pi/2 - (\varphi_e - D_e)) + (Z_e + C_e)^2 \tag{5}$$

$$F_i^2 = H_i^2 + B_i^2 + 2HB\,(\pi/2 + (\varphi_i - D_i)) + (Z_i + C_i)^2 \tag{6}$$

We now suppose that the measurements are performed in a large vertical coil by means of which the ambient vertical field is annulled as it is done when the vector method is used for absolute measurement of H (Kring Lauridsen, 1980). We also assume that the D-coil is properly levelled so that its vertical component C is negligible.

The D-coil is provided with a telescope the axis of which is horizontal and nearly perpendicular to the axis of the coil. In each position this telescope is pointed at a mark which is placed with good approximation in the magnetic meridian through the point of observation, either north or south of the pier. The inversion is a 180° rotation of the coil about the axis of the telescope and as illustrated in Figure 2 we have that

$$\varphi_e = a_P + c; \qquad \varphi_i = a_P - c, \tag{7}$$

where a_P is the azimuth of the mark P and c is the collimation angle between the axis of the telescope and the axis of the coil. The angle c is approximately 90° and with this arrangement we always assume that the mean direction of the coil axis in the two positions is the direction to the mark and that the direction of the axis of the coil is in accordance with (3).

It is useful to introduce the following notation for the mean (or half sum) and the half difference in the erect and inverted positions:

$$\langle A \rangle = \tfrac{1}{2}(A_e + A_i); \qquad [A] = \tfrac{1}{2}(A_e - A_i). \tag{8}$$

Further let

$$E_e = \frac{F_e^2 - B_e^2 - H_e^2}{2HB}; \qquad E_i = \frac{F_i^2 - B_i^2 - H_i^2}{2HB} \tag{9}$$

then (5), (6), (7) result in the formulas

$$\langle D \rangle = a_P + [E] \tag{10}$$

$$\pi/2 - c = \langle E \rangle - \langle D \rangle. \tag{11}$$

The formula for $\langle D \rangle$ can also be written in another way. For the square of a quantity

$$[A^2] = 2\langle A \rangle [A]$$

and using this for the terms in E we get explicitly

$$\langle D \rangle = a_P + \frac{F(F_e - F_i)}{2HB} - \frac{B_e - B_i}{2H} - \frac{H_e - H_i}{2B}. \tag{12}$$

From this formula it appears that the absolute values of B, H, and F are not important for the measurement of D. It is first and foremost the variation of these quantities during the readings in erect and inverted position that matters. The elements D and H are derived from variometers or a fluxgate system. F is read directly from the proton magnetometer and in order to keep record of B it is necessary to measure the current in the coil accurately. The obvious solution is a standard resistor the voltage across which is measured by means of a digital voltmeter. This system is also useful for the compensation method (Kring Lauridsen, 1980). If the scale value of the coil is S_U nT per unit of the digital voltmeter then $B = S_U U$.

Till now the magnitude of the coil field B has remained an open question. From (12) it appears that the larger B is the better is the resolution. On the other hand the imperfection of the coil field homogeneity can put a limit to the applied value of B.

3. Practical Details and Experimental Results

From the considerations above it is evident that it is essential for the present method to have access to a manageable coil with sufficiently homogeneous field B for deviation of the ambient horizontal field vector. Luckily a coil system fulfilling these needs has recently been proposed by Robinson (1983).

This coil system consists of 4 separate coaxial windings with a common diameter. The two inner sets of windings have 4 turns each, and the outer sets have 9 turns each, or possibly another similar number of turns with the same ratio. Of course the distances between the windings depends on the diameter but also to a certain degree on the wanted degree of homogeneity. For details the reader is referred to the original paper by Robinson.

Our first intention was to wind the coil on a drain-pipe made from PVC, but this tube proved to be far from circular. Then a tube made from paper laminated bakelite was ordered from a Danish factory and this tube has a very good geometry. The inner diameter is 300 mm, the wall thickness is 8 mm and the length is 320 mm. Grooves for the windings were turned on a lathe.

The coil is placed on four supporting pieces fixed on a turntable which is placed directly on the marble plate of the observation pier and is provided with a fine adjustment. In this way measurements of D can be made without resorting to a theodolite. It goes without saying that the azimuth of the mark cannot be determined without a theodolite but in principle it can be done by means of an ordinary (magnetic) survey theodolite. The reason for the above arrangement is that it is easier to obtain a suitable height for the point of reference (or rather volume of reference because the proton liquid has a certain extension). If the coil was placed on top of a theodolite the reference point would either be high above the pier or a rather large and unmanageable coil surrounding pier and theodolite would be needed.

In the middle of the coil tube two holes are made. A telescope is introduced in one hole and firmly fixed to the tube looking through the other hole towards the mark. The proton magnetometer sensor must of course be placed in the centre of the coil and therefore it cannot be introduced till the mark has been sighted at. It is important that the direction of the coil does not change when the sensor is introduced. Therefore the cylindrical sensor is guided but an U-shaped PVC duct which is supported independently of the coil. Figures 4 to 7 illustrate the arrangement.

To sum up the measurements are performed in the following way. It is necessary to

Fig. 4. The compensation coils at Brorfelde. The *Z*-coil is so large that there is easy access to the observation pier. On the photo a QD is placed on the pier. The *H*-coil is fastened to the ceiling. Its lower frames pass below the floor.

Fig. 5. The proton sensor is guided by a *U*-shaped PVC duct (for cable mounting).

Fig. 6. The telescope is fastened to the cylinder by means of a plate and looking through an opposite hole.

Fig. 7. The *D*-coil rests on a circular disc which is provided with a fine adjustment. The two supports for the coil and for the sensor are independent of one another.

have a mark nearly in the magnetic meridian, either north of or south of the pier. But this mark can have a provisional character if its azimuth is controlled by means of a standard mark in connexion with the measurements.

First the ambient total field on the pier is measured by means of the proton magnetometer. Then the vertical component Z is turned upside down by means of a suitable current in the large compensation coil. The corresponding value of the current is halved and thereby $Z = 0$ is obtained as described in detail elsewhere (Kring Lauridsen, 1980). Now the D-coil is placed on the turntable exactly centered on the pier and the telescope is pointed at the mark. The current in the coil is switched on with the B-field pointing towards east. The proton magnetometer sensor is placed in the centre of the D-coil and a series of readings are taken comprising F, U (from the digital voltmeter), D- and H-variometer. Next the sensor is removed, the D-coil inverted, the sensor replaced and another series of readings are taken with the B-field west.

D is now calculated from (10) or (12). It is also possible to calculate the collimation angle, or rather $\pi/2$-c from (11) as a check. This quantity should of course remain nearly constant.

The magnitude of the coil field B is to a certain degree arbitrary. In principle it should be as large as possible, but in practice it will probably be found that when a certain limit is exceeded then the proton magnetometer no longer gives consistent readings due to inhomogeneities in the coil field. At Brorfelde we found that $B \approx H \approx 17\,000$ nT was a suitable value.

In order to obtain a slow drift in the B-field during the readings it is recommended that an extra resistor is used in series with the D-coil if this has a low ohmic resistance. We used a resistor of the make Vishay which has an extremely small variation of resistance with temperature. These resistors are therefore also useful as standard resistors for measurements of current with a digital voltmeter.

If the Z-field of the compensation coil is not lying exactly in the meridian plane an error is introduced because the value of D will differ from the ambient field. The corresponding problem is met with when H is measured with the compensation method (Kring Lauridsen, 1980) and the solution is quite similar. The change in D when the compensation current is switched on in the Z-coil is observed by means of a 'differential variometer' with liquid damping or by means of a QD which also has a heavy damping of the suspended magnet. Either the change in D is corrected for, or the Z-coil is adjusted until no change is observed during the transition from natural field to compensated field $Z = 0$.

Another source of error is magnetism in the proton sensor. Mr. A. Ruta has pointed out that this error can be avoided by reversing the sensor when the D-coil is inverted.

The method described above is based on the assumption that Z is nearly compensated. It is natural to ask the question whether this compensation is necessary. It depends of course on the numerical value of Z in the ambient field. If $|Z|$ is large some difficulties turn up. In (5) and (6) we can no longer neglect the last term $(Z + C)^2 = Z^2 + C^2 + 2ZC$. This means that firstly we have to keep record of variations of Z on top of those of D and H. Furthermore C becomes critical because the term $2ZC$ is no longer of second order. This could be counteracted by reversing the current in the D-coil both in the erect and the inverted position causing a change of the sign of $2ZC$. But this results in a doubling of the measurement scheme and more complicated formulas. C is strictly proportional to the current in each position but the proportionality factor is not necessarily the same in the erect and the inverted position when the demands of accuracy are high due to the term $2ZC$.

The D-variometer used in our case is a la Cour variometer put up in the absolute house for direct eye-reading. The scale value is $1'\,\mathrm{mm}^{-1}$ and the readings are made in tenths of minutes of arc. This implies subjective estimation of tenths of division which introduces a certain source of error. All readings are made by the same assistant and from an analysis of the frequency distribution of a total of 2000 readings along the ideas of H. Bäckström (1930, 1932) it is concluded that a single reading has a standard error of about 0.1′ exclusively dating from the subjective estimation of tenths of divisions. A similar analysis of the readings of the theodolite used for QD-measurements yielded a standard error of the same order of magnitude. The theodolite is a home made instrument with scale microscopes each estimated to 1/20 of double minutes of arc.

Table I and II outlines the results of some measurements of D at Brorfelde (BFE) with

TABLE I

Base-line value D_0 for D-variometer
Mean of N observations BFE 1983

(June–July)

Observatory	Instrument	D_0	Standard error	N
BFE	D-coil (large sensor)	−81.8′	0.19′	10
	QD 4	82.1	0.13	11
	QD 5	82.1	0.20	12
LOV	Askania declinometer	82.1		3
	Elsec DI-flux	82.1		4
NUR	Askania Torsion declinometer	82.4	0.29	7
	QD 8	82.2	0.23	6
TRO	QD 10	82.2		4
LRV	QD 11	82.6		3

TABLE II

(November)

Observatory	Instrument	D_0	Standard error	N
BFE	D-coil (small sensor)	$-81.5_5'$	0.11′	9
	D-coil (large sensor)	$81.6_5'$	0.06	8
	QD 4	81.6_5	0.10	9
	QD 5	81.7	0.14	8
CLF	DI-flux	81.6_5	0.14	5

the D-coil and some other methods. The first three lines of Table I refers to instruments belonging to the BFE observatory, while the rest of the table refers to measurements made during the yearly Nordic comparison meeting.

The sensor which hitherto had been used with the proton magnetometer was of the double coil type. As this sensor is of rather large dimensions relative to the D-coil it was possible that a smaller sensor would be of advantage. Such a sensor of the single coil type was ordered from Elsec which makes them for bore hole logging. The sensor we received is slightly magnetic.

Table II outlines the results of some measurements made during October–December 1983 with the small and the large sensor and other methods. During this period there was a slight drift in the base-line value D_0 ($0.005' \text{day}^{-1}$) and in order to compare the measurements they were reduced to the beginning of November. At this time the observatory was visited by D. Gilbert from Institut de Physique du Globe, Paris, who carried the DI-flux developped in France. He made comparison observations of all elements.

A measurement with the D-coil consists of 2×10 readings with intervals of $\frac{1}{2}$ min, while a measurement with QD consists of 2×5 readings with intervals of 1 min. Thus the total time used for a measurement is nearly the same for the two methods. However, the greater number of readings with the D-coil is reflected in corresponding smaller standard error.

From Table II it appears that there is no significant difference between the results with the two sensors, and the systematic difference between the D-coil and the QD's has disappeared. There is no significant difference between the values obtained by the different instruments.

An outlier with QD 5 and one with D-coil, each deviating about 0.5′, have been discarded. Both QD 4 and QD 5 have been used with the telescope belonging to the theodolite in order to attain maximum sensivity.

References

Bäckström, H.: 1930, 'Über die Dezimalgleichung beim Ablesen von Skalen', *Zeitschr. f. Instrumentenkunde* **50**, 561–575, 609–624.

Bäckström, H.: 1932, 'Über die Dezimalgleichung beim Ablesen van Skalen', *Zeitschr. f. Instrumentenkunde* **52**, 105–123, 260–274.

Kring Lauridsen, E.: 'QD. A Quartz Declinometer of Classical Type', Danish Meteorological Institute, Geophysical Papers R-65. København 1981.

Kring Lauridsen, E.: Absolute Measurements with Proton Magnetometers in Stationary Coils. Danish Meteorological Institute, Geophysical Papers R-57, København 1980.

Robinson, P. R.: 1983, 'Improvements to the System of Four Equiradial Coils for Producing a Uniform Magnetic Field', *J. Phys. E: Sci. Instrum.* **16**, 39–42.

PORTABLE MAGNETOMETER THEODOLITE WITH FLUXGATE SENSOR FOR EARTH'S MAGNETIC FIELD COMPONENT MEASUREMENTS

J. BITTERLY, J. M. CANTIN, R. SCHLICH, J. FOLQUES
Institut de Physique du Globe, Université Louis Pasteur, 5 rue René Descartes, 67084 Strasbourg Cedex, France

and

D. GILBERT
Institut de Physique du Globe, Université P. et M. Curie, 4 place Jussieu, 75230 Paris Cedex 05, France

Abstract. A new portable magnetometer theodolite with fluxgate sensor has been developed for measuring the components of the earth's magnetic field. With this instrument, used as a null detector, it is possible to make absolute measurements of declination (D) and inclination (I). The compensation of the main magnetic field allows direct intensity measurements (H, X, Y, Z components); in this case the calibration is made by reference to a proton magnetometer.

The observation procedure is particularly easy and quick. The accuracy is better than 5 sec of arc for D and I and better than 1 nT for direct intensity measurements.

1. Introduction

In the French subantarctic magnetic observatories, declination (D) and inclination (I) measurements of the earth's magnetic field are made routinely with a fluxgate sensor mounted parallel to the optical axis of the ZEISS 010 B theodolite telescope; the sensor is used as a null detector. The sensor offset and the relative misalignment between the fluxgate sensor and the telescope are both cancelled by measurement processes.

Magnetometers using this measurements method has been described by several authors since 1954 (Meyer and Voppel, 1954; Serson and Hannaford, 1956; Trigg, 1970).

However their stability and their accuracy were insufficient for absolute determinations in the magnetic observatories.

A new portable magnetometer was developed in 1978 by the Institut de Physique du Globe; it has now been adapted to allow direct intensity measurements in any directions. Intensity measurements imply stable compensation of the earth's main magnetic field. The precise value of the compensated field and the residual field detected by the sensor allow the computation of the field intensity in a given direction. The compensation of the main field is obtained by an original, stable and super-linear current generator. The calibration of the compensation circuit is made by reference to a proton magnetometer (F, total field intensity measurement). This calibration is part of the measurement procedure which cancels simultaneously the instrumental errors by turning round and over the theodolite and the fluxgate sensor. Direct measurements in nanotesla (nT) of the magnetic field components (H, X, Y, or Z) are particularly easy

Geophysical Surveys **6** (1984) 233-239. 0046-5763/84/0063-0233$01.05.

and quick (5 min per component). Six magnetometers have been built and are now used at the permanent magnetic observatories of Chambon-la-Forêt in France, Port-Alfred, Port-aux-Français, Martin-de-Viviès in the Indian Ocean and Dumont d'Urville in the Antarctic.

2. General Description

The principal characteristics of this portable magnetometer theodolite are given below:

– The theodolite is a ZEISS JENA 010 B specially adapted for this application (mean error: $\pm 1''$ horizontal and vertical circles; level sensitivity: $1''$ estimated).

– The fluxgate sensor has been developed by THOMSON-CSF (length 50 mm, diameter 18 mm). The output sensitivity is of 2 nV nT^{-1}, the output dynamic range is plus/minus 1000 nT, the noise is of 0.2 nT peak to peak from DC to 0.5 Hz, the temperature sensitivity of the sensing head is less than 0.05 nT $°C^{-1}$ in zero magnetic field and the fluxgate sensor offset is less than 2 nT.

– Electronics and controls, built by 'Institut de Physique du Globe de Strasbourg', are mounted in an aluminium carrying case. The following specifications have been obtained:

- power consumption: 100 mA/12 V (Pb sealed battery 12 V/4 Ah),
- resolution (null detector): 0.1 nT; dynamic range: ± 350 nT full scale,
- stability of compensation current: better than 2 p.p.m $°C^{-1}$,
- linearity of constant current generator: better than 10^{-5},
- temperature sensitivity of electronic housing: 0.2 nT $°C^{-1}$ for 50 000 nT compensated field,
- accessories for field measurements are available (non magnetic reading box with time and null detector displays, non magnetic tripod).

The schematic circuit diagram is given on Figure 1. For this instrument the fluxgate sensor offset is very low (from minus 2 nT to plus 2 nT) and keeps steady for a long period of time. For example the average sensor offset, computed for 170 measurements at the magnetic observatory of Martin-de-Viviès (Amsterdam Island) during a period of 12 months without any adjustment, is equal to 0.6 nT (standard deviation: 0.8 nT).

3. Comparison with standards

In the years 1979–1980, 127 measurements of declination and inclination have been compared to the standards given by the magnetic observatory of Chambon-la-Forêt. The results show very good agreement between the portable magnetometer and these standards; no apparent long term drift has been observed. The average differences between the portable magnetometer theodolite data and the corresponding standards data are the following:

$$\overline{\Delta D} = -11'' \text{ (r.m.s. deviation} = 20'');$$
$$\overline{\Delta Z} = -0.3\,\text{nT (r.m.s. deviation} = 0.8\,\text{nT});$$
$$\overline{\Delta H} = +1.0\,\text{nT (r.m.s. deviation} = 1.3\,\text{nT}).$$

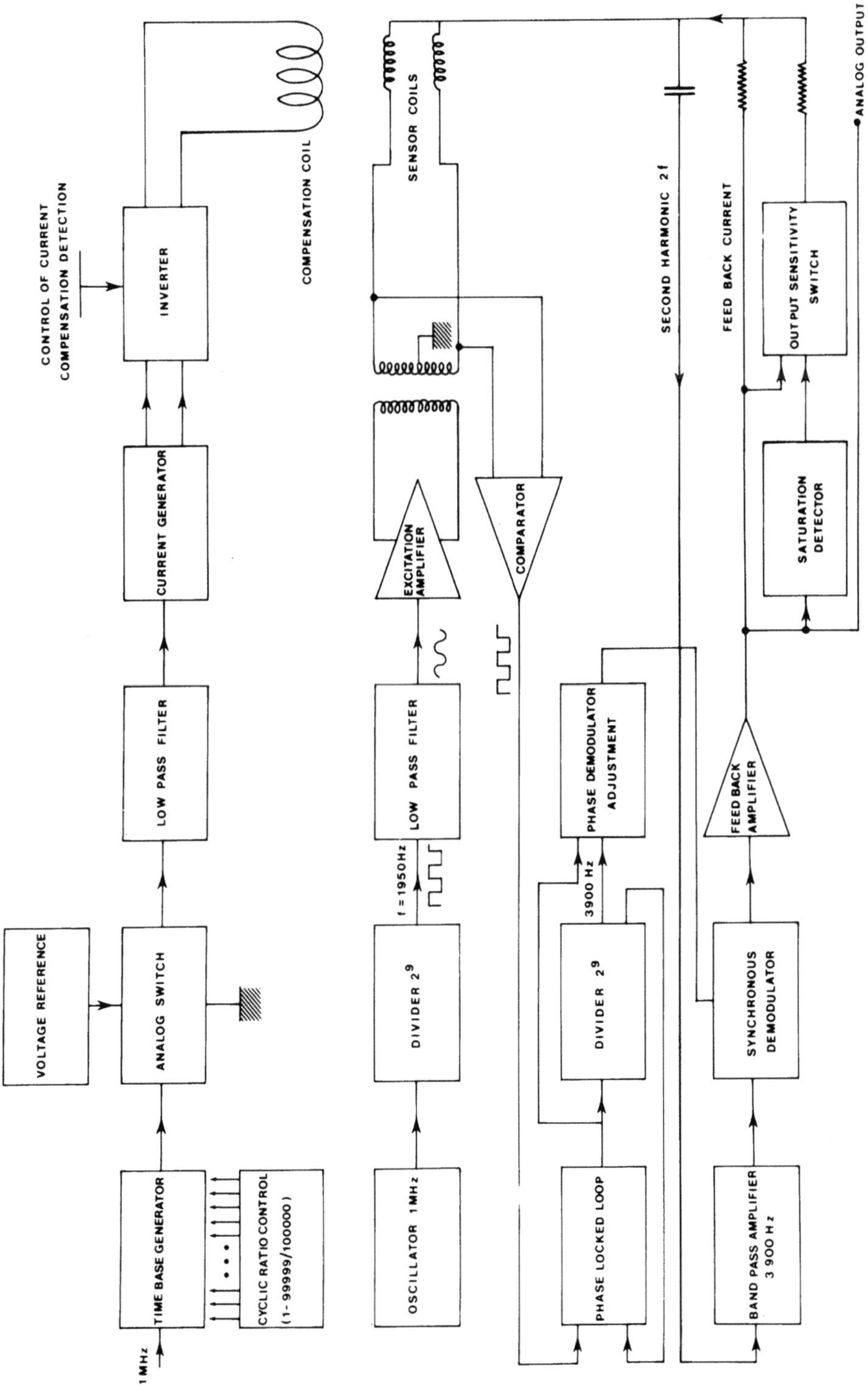

Fig. 1. Schematic circuit diagram of the portable magnetometer theodolite.

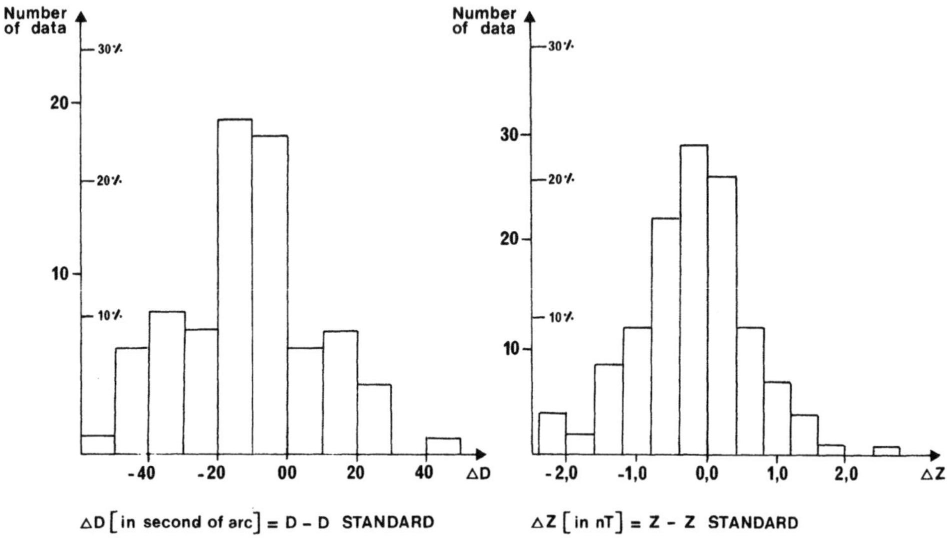

Fig. 2. Distribution of the difference between the measurements made with portable magnetometer theodolite and the measurements made with the observatory standards at Chambon-la-Forêt. ΔD for declination measurements ΔZ for Z values computed from I and F measurements

Figure 2 shows these results for declination and vertical component measurements.

At the magnetic observatory of Port-Alfred (Crozet Island) the analysis of 91 measurements (inclination measurements with a portable magnetometer theodolite and H measurements with the Q.H.M. 732) has shown, for H determinations, an agreement in 94% of cases within ± 1.0 nT.

At the magnetic observatory of Port-aux-Français (Kerguelen Island) the variometer base line values computed from the magnetometer theodolite measurements (D, I) or from the magnetic observatory standards (Chasselon theodolite No. 66501, Q.H.M. No. 726) confirm the results described above.

4. Dispersion and Accuracy of Results

For this study 508 measurements made in the subantarctic French magnetic observatories have been taken into account (Kerguelen: 224, Crozet: 216, Amsterdam Island: 68). The dispersion is less than 5 sec of arc for D measurements and remains smaller than ± 1 nT for H and Z values computed from I and F measurements. For direct component measurements (H and Z) the results are concordant and present the same dispersion. As Figure 3 shows, the values computed for H and Z components, by either the null method or by the compensation method, are identical within ± 1 nT in 84% of the cases.

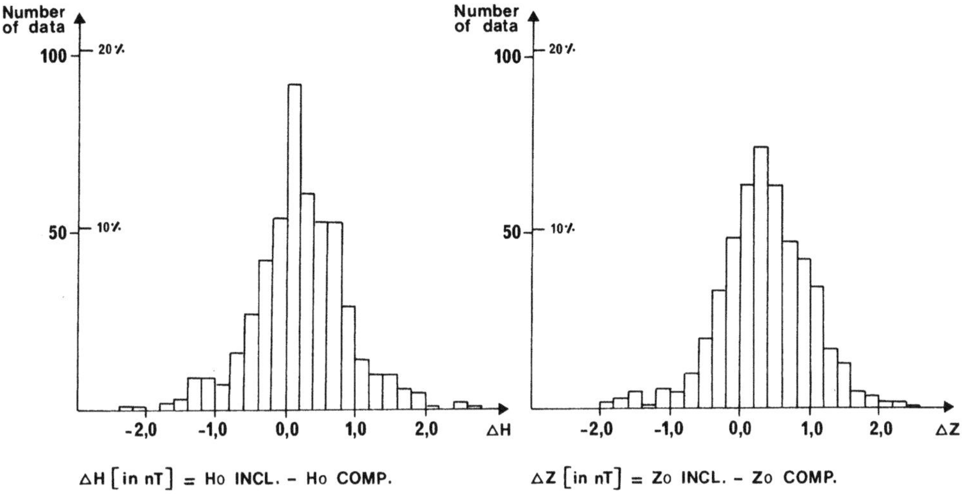

Fig. 3. Distribution of the difference between the base line values H_0 and Z_0 computed from inclination measurements (null method) and from components (*H* and *Z*) direct measurements (compensation method).

5. Direct Measurements of *X*, *H*, *Y*, *Z* Components in Polar Regions

With this instrument, the direct measurements of *X*, *Y*, *H*, or *Z* components for values ranging between 0 and 80000 nT remain possible during disturbed magnetic conditions, since the measurement procedure does not involve a null field position which is the case for *D* and *I* measurements. This facility is greatly appreciated in polar regions for weak horizontal field measurements where the Q.H.M. method appears often very delicate. Figure 4 shows an example of *X*, *Y*, *Z* determinations by the compensation method in disturbed magnetic conditions at the magnetic observatory of Dumont d'Urville in Antarctica. At Dumont d'Urville the X and Y horizontal components values are respectively 1000 nT and 450 nT; the vertical component's value is 70 100 nT. The base line values (X_0, Y_0, Z_0) computed from the absolute determinations (*X*, *Y*, *Z* direct measurements) made at time 1 and 2 are in excellent accordance in spite of the 100 nT earth's magnetic field variation recorded between time 1 and 2.

6. Conclusions

The portable magnetometer theodolite allows easy and quick measurements of the earth's magnetic field. The accuracy is better than 5 sec of arc for *D* and *I* measurements and better than 1 nT for intensity measurements (*H*, *X*, *Y*, *Z*). Considering these results it is obvious that the portable magnetometer theodolite can be proposed as a standard for magnetic observatories. This instrument is also appreciated for repeat station measurements and magnetic field surveys work.

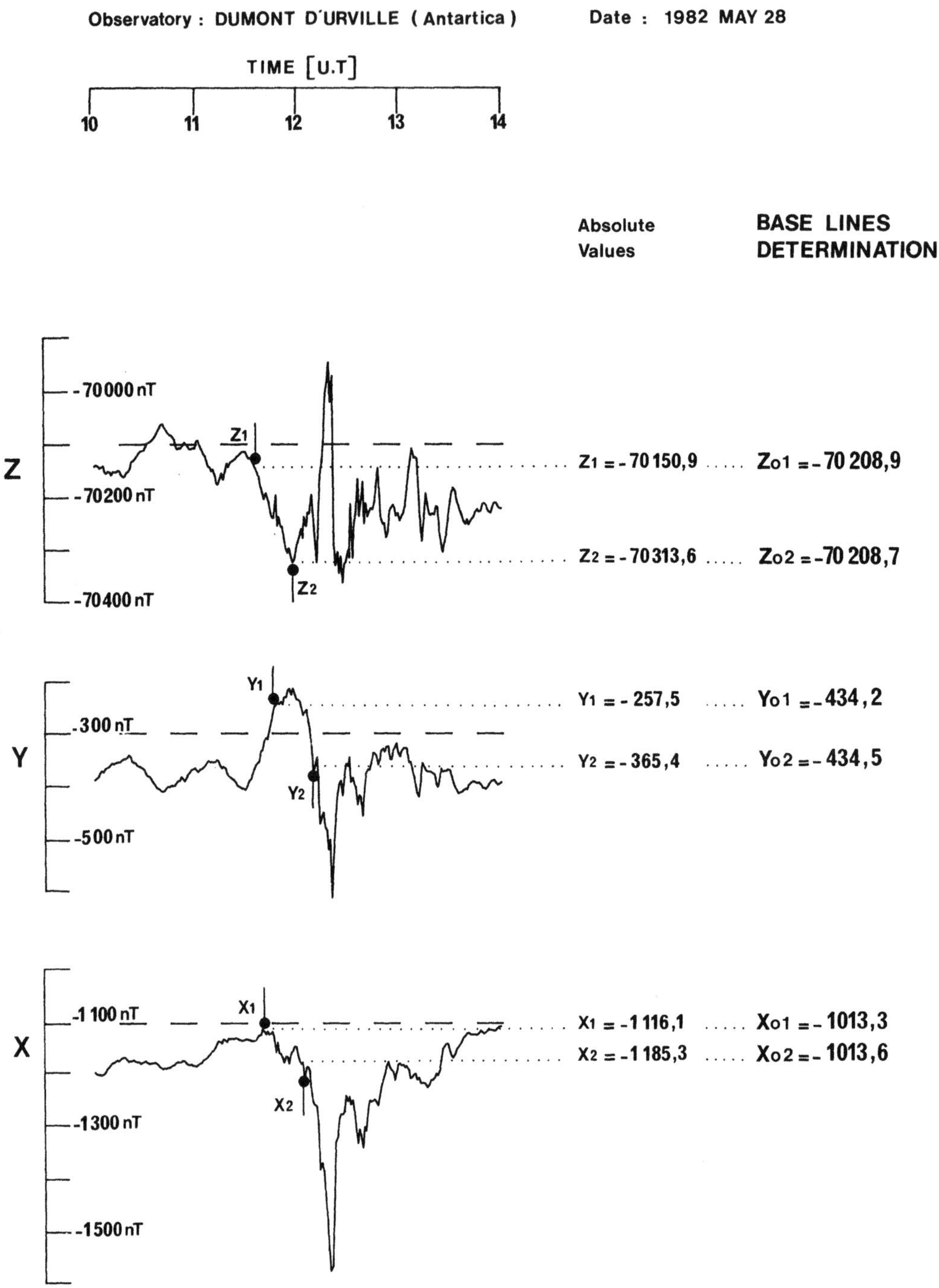

Fig. 4. Example of X, Y, Z determinations by the compensation method in disturbed magnetic conditions at the magnetic observatory of Dumont d'Urville (Antarctica).

References

Meyer, O. und Voppel, D.: 1954, 'Ein Theodolit zur Messung des erdmagnetischen Feldes mit der Förstersonde als Nullfeldindikator', *Deutsche Hydrographische Z.* **7**, 73–77.
Serson, P. H. and Hannaford, W. L. W.: 1956, 'A Portable Electrical Magnetometer', *Can. J. Technol.* **34**, 232–243.
Trigg, D. F.: 1970, *A Portable D and I Magnetometer*, Geomagnetic Lab. Report No. 70–3, Earth Phys. Branch, Dept. Energy, Mines and Resources, Ottawa, Canada.

ACQUISITION AND PRIMARY PROCESSING OF GEOMAGNETIC PULSATIONS IN REAL TIME OPERATION BY MICROCOMPUTER

H.-J. LINTHE and D. LENNERS

Central Institute of Solar-Terrestrial Physics, Acad. Sci. G.D.R., G.D.R. 1199, Berlin

Abstract. For continuous acquisition and primary processing of short periodic geomagnetic variations at the Adolf Schmidt Observatory Niemegk recording equipment has been put into operation which renders possible a permanent computer controlled observation of pulsations in the framework of geomagnetic observatory tasks. It permits the stepwise replacement of processing analogue paper recordings practised for several years and extends the on line recordings with the process control computer system PRS 4000 carried out up to now on the monthly 3 World Days only. In addition to routine characterization of pulsation activity, especially by use of process control computers for on line primary processing extended investigations on the microstructure of diurnal variations of pc pulsation amplitudes in medium latitudes became possible. Now, these investigations can be significantly enlarged in a permanent observation with problem oriented on line data condensation.

For continuous acquisition and primary processing of short-periodic geomagnetic variations at the Adolf Schmidt Observatory Niemegk recording equipment has been put into operation which renders possible a permanent computer-controlled observation of pulsations in the framework of geomagnetic observatory tasks. It permits the stepwise replacement of processing analogue paper recordings practised for several years and extends the on-line recordings with the process control computer system PRS 4000 carried out up to now on the monthly 3 World Days only. In addition to routine characterization of pulsation activity, especially by use of process control computers for on-line primary processing, extended investigations on the microstructure of diurnal variations of pc pulsation amplitudes in medium latitudes became possible. Now, these investigations can be significantly enlarged in a permanent observation with problem-oriented on-line data condensation.

The intention of on-line primary processing carried out up to now with the process control computer was the formation of 6-min short-time mean values of amplitude and period, in the period ranges pc 2 to pc 5, of horizontal components X and Y of geomagnetic pulsations. Table I shows the limiting periods of the four standard period bands pc 2, 3, 4, and 5.
By use of digital band pass filtration with following determination of spectral mean values, a strong data condensation with simultaneously greatest possible information content of the mean values for the 4 period ranges from 5 to 600 s can be reached. Inherent in the mean value calculation is the supposition that the period of the presumed sinusoidal time function does not vary too strongly in the interval in which the mean period value is to be determined.

Because of the many years of experience in the use of the process control computer

Geophysical Surveys **6** (1984) 241-247. 0046-5763/84/0063-0241$01.05.

TABLE I

Limiting values of the period bands pc 2, 3, 4, and 5

pc	limiting values s	Hz	band-width /Hz	center frequency /Hz
2	5...10	0.2 ...0.1	0.1	0.15
3	10... 45	0.1 ...0.022	0.078	0.061
4	45...150	0.022 ...0.0066	0.0154	0.0143
5	150...600	0.0066...0.0017	0.0049	0.00415

PRS 4000 for real-time observation of short-period variations, this kind of problem-oriented primary processing was used for the application of the microcomputer also. Contrary to the zero phase filter used up to now which actually provides band passes with a high slope steepness, but is very computing intensive because of the great number of multiplications, new algorithms, adjusted to the operating speed of the microcomputer, had to be selected for the band pass filters.

Concerning the filter design, certain characteristics have to be considered which are realizable with the existing technical possibilities as good as possible. These are limiting frequency and centre frequency, slope steepness, pass band and stop band attenuation and phase characteristic. Well-known methods which approach an ideal pass band curve are Butterworth, Tschebyscheff, and Cauer filters. Butterworth band pass filters actually have a smaller slope steepness compared with the other but no overshoot in the pass band and stop band range. That they have the mathematically simplest approximation function is a further significant advantage. Because of this, the Butterworth band pass filtration was chosen for the present application. A sufficient slope steepness was guaranteed by use of filters of 6-th order. With Butterworth filters, as with other recursive filter types too, one has to consider that the phase relation of the output compared with that of the input is out of phase. A compensation for the phase displacement was reached by applying the filter algorithm first to the input time series, then applying the same algorithm in reverse direction to the output time series of the forward filtering. The resultant time series calculated then has the same phase as the input time series.

In selecting an appropriate microcomputer a modular system with universal input and output components and a big memory space was opted for. Figure 1 shows the different functional units of the computer system as they were used for the present recording system. Though acquisition and primary processing of short-period variations are in the foreground, the recording system presented in figure 1 can process data massifs of smaller volume (interrogating in minute intervals) too. Additionally, a computer coupling to the superordinate process control computer exists which organizes long-term storage and processing of the results of primary processing. The central processing unit is fitted with the 8-bit microprocessor Z 80 (Figure 1). For semiconductor memories, 4 K-byte EPROM and 20 K byte RAM are used which can

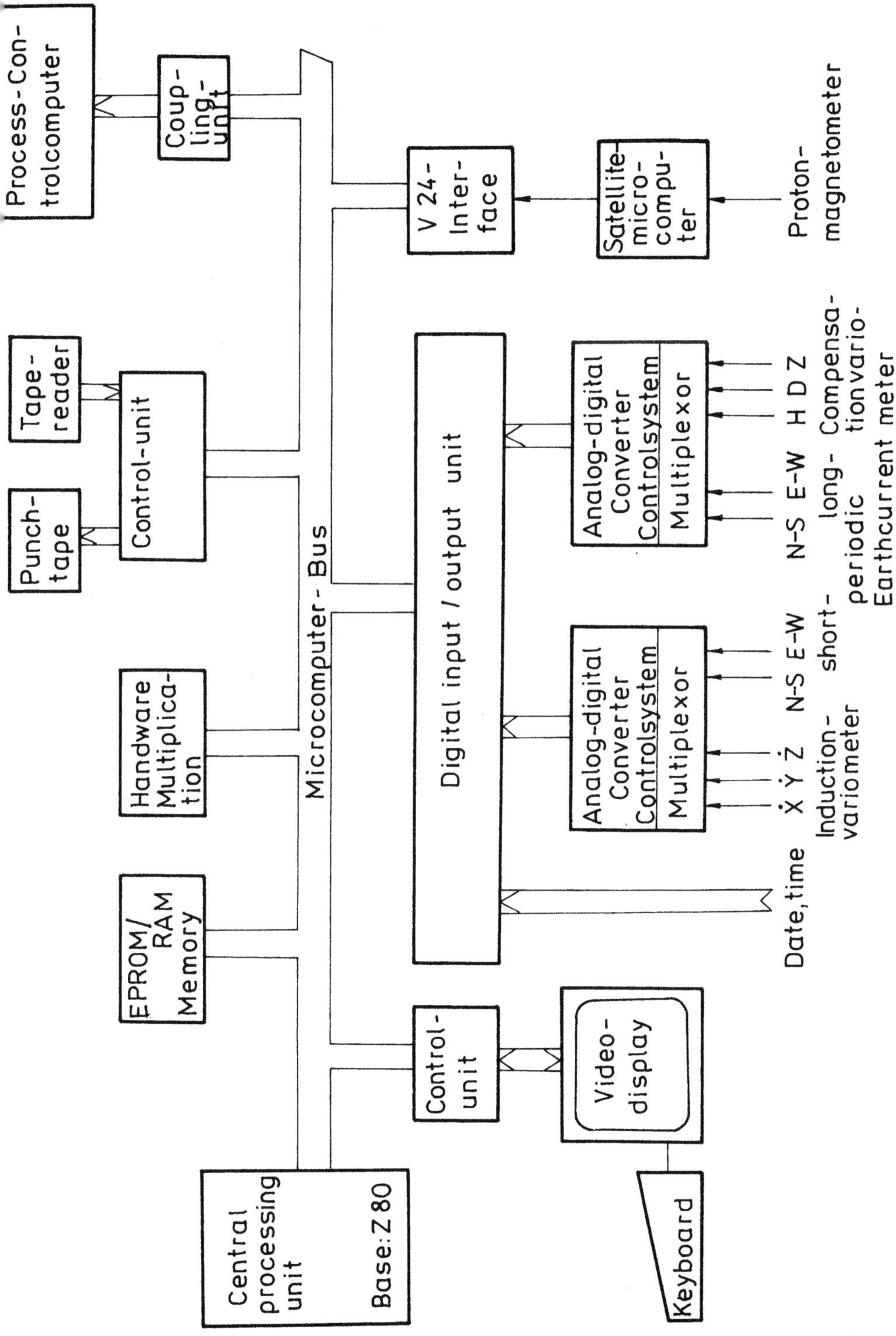

Fig. 1. Microcomputer controlled recording system.

be extended in modular way up to 64 K-byte. A hardware multiplication unit supports the software and contributes thus to an increased processing speed.

The punched tape input is mainly necessary for reading in of variable scale values. The punched tape output is one of the output possibilities for results of the on-line primary processing. The output of result by computer coupling with the superordinate process control computer, however, is of higher priority. For operator communication, an alphanumeric video display with input keyboard is provided.

Process coupling is mainly achieved by means of the digital input unit. Date (year, day of the year) and time (hr. min, sec) are made available by the central master clock system of data acquisition. The geomagnetic pulsation voltages X, Y, and Z measured by the induction variometers are – supplemented by the geoelectric pulsation of the N–S and E–W component – detected by a fast multiplexor, digitized and sequence checked and passed on in series in 2-byte representation to the digital input.

Similarly the acquisition of long-period geomagnetic components H, D, and Z and long-period geoelectric components of N–S and E–W direction is carried out, the data volume of which however, is significantly smaller than that of the short-period variations because of the 1-min sampling.

A further process coupling exists with the V24 Interface by means of which the coupling to a subordinated microcomputer is realized. It forms the base in computing technique for a coupled measuring system consisting of proton magnetometer and ferro-probes by means of which permanent recording of long-period geomagnetic variations measured in 1-min intervals yields measured values which are checked for veracity.

The most important aspects of the mode of operation of the recording system are to be seen in the representation of software components in Figure 2. By interrupt, the adjustment of initial conditions is given with the start of the system. The real-time clock is software realized and provides date and time information for the measured value files. It is timed with the 1-sec interrupt and after adjustment, time-cyclically compared with the data of the central master clock system at the moment of start. In the acquisition of measured values for 1-sec samples the digitized values of the short-period variations are taken by interrupt-dependent system and compiled in files. Measured value error detection and other error measures derived from this, guarantee further processing only of measured values which have been verified. Analogously the acquisition of long-period variation data with a 1-min sampling interval and the compilation into their own result file is carried out. The selection process which eliminates faulty measurements of the 1-min sampling together with the base of dates supplied and checked by the satellite computer, generates a measured value file which can be called by the computer coupling from the superordinate process control computer.

The digital filtration is performed on the measured value file of the 1-sec samples only. As already described above, after a digital band pass filtering of pulsations in the period ranges pc 2 to pc 5, the spectral 6-min mean values are determined for period and amplitude. The band pass filters are realized by recursive Butterworth techniques. For suppression of their non-linear phase shift the filtration is carried out in forward and

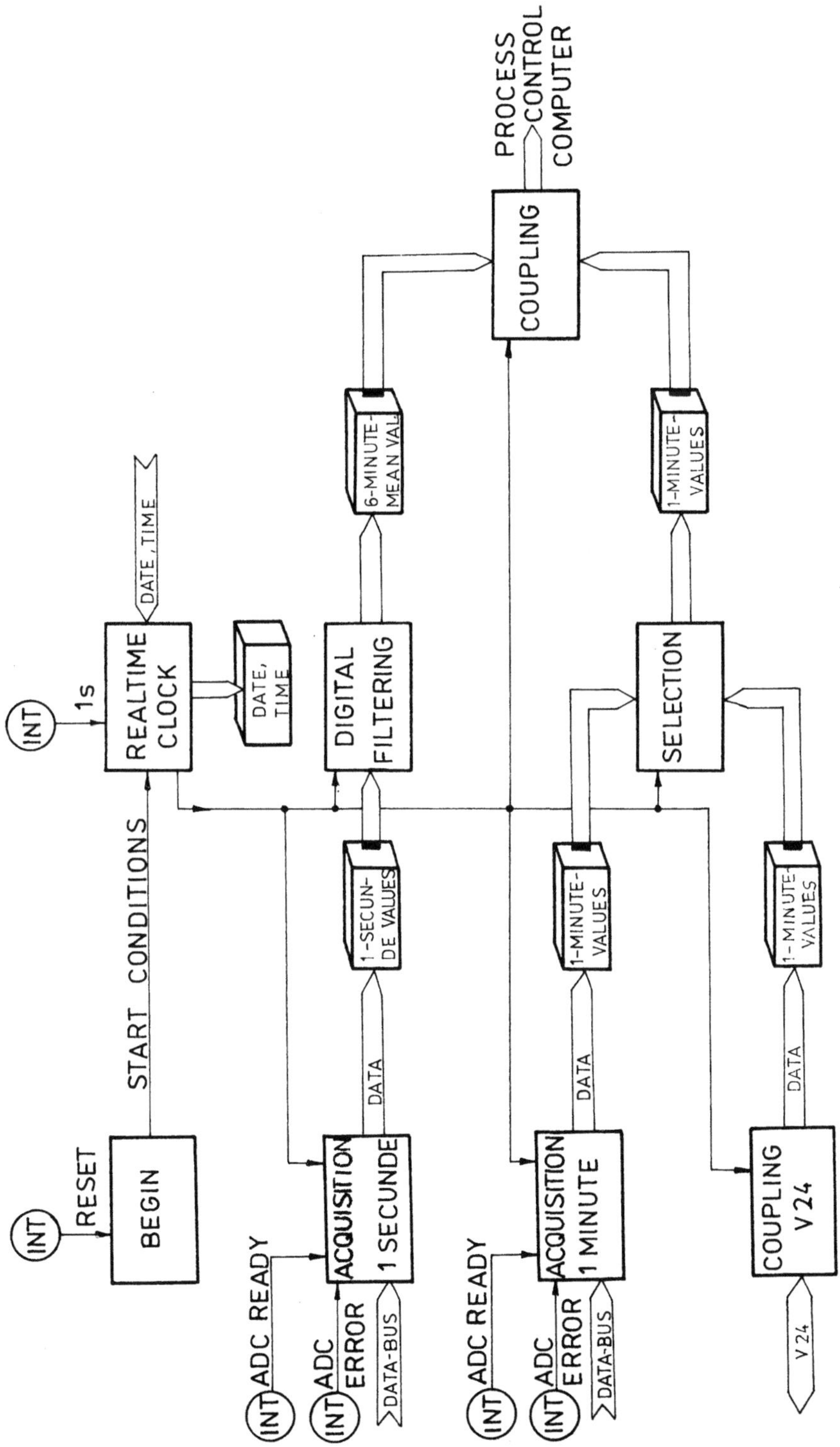

Fig. 2. Software-components of the recording system.

TABLE II

Example of output of the microcomputer controlled recording system

	X ampl	X per	Y ampl	Y per	EY ampl	EY per	
10.2.84 22.00	0.00	0.00	0.00	0.00	0.00	0.00	pc2
	0.00	0.00	0.00	0.00	0.40	25.09	pc3
	0.41	86.05	0.00	0.00	0.00	0.00	pc4
	1.31	230.56	1.44	271.31	0.00	0.00	pc5
10. 2.84 22.06	0.00	0.00	0.00	0.00	0.00	0.00	pc2
	0.13	22.98	0.12	27.79	0.39	23.36	pc3
	0.70	81.35	0.33	70.99	1.22	65.16	pc4
10. 2.84 22.12	0.00	0.00	0.00	0.00	0.00	0.00	
	0.46	27.71	0.21	24.55	1.30	27.56	
	1.16	75.17	0.86	71.14	3.16	68.02	
	1.94	266.52	2.38	308.88	5.65	273.23	
10. 2.84 22.18	0.00	0.00	0.00	0.00	0.00	0.00	
	0.55	31.05	0.46	29.83	1.59	31.88	
	0.46	63.77	0.77	60.94	1.45	60.80	
10. 2.84 22.24	0.00	0.00	0.00	0.00	0.00	0.00	
	0.13	24.29	0.14	25.18	0.96	27.53	
	0.40	59.05	0.72	71.72	2.50	77.94	
	0.69	198.06	1.43	209.44	5.28	222.27	
10. 2.84 22.30	0.06	8.85	0.06	9.37	0.48	8.61	
	0.45	23.37	0.65	26.97	2.16	22.66	
	1.27	63.73	1.56	59.07	5.10	77.21	
10. 2.84 22.36	0.05	8.78	0.00	0.00	0.34	8.18	
	0.36	23.44	0.34	23.87	1.45	23.23	
	0.98	71.44	1.45	69.64	3.35	63.75	
	1.27	252.83	1.16	197.41	6.38	343.31	
10. 2.84 22.42	0.00	0.00	0.00	0.00	0.00	0.00	
	0.17	20.65	0.14	21.03	0.88	23.46	
	0.51	71.68	0.77	66.63	2.82	69.36	
10. 2.84 22.48	0.00	0.00	0.00	0.00	0.00	0.00	
	0.13	26.02	0.16	29.93	0.49	25.23	
	0.50	70.59	0.60	70.62	1.51	70.59	
	1.38	241.19	0.94	217.95	4.90	269.13	
10. 2.84 22.54	0.00	0.00	0.00	0.00	0.00	0.00	
	0.11	26.09	0.11	26.37	0.38	24.18	
	0.45	65.66	0.63	81.45	1.02	68.64	
10. 2.84 23.00	0.00	0.00	0.00	0.00	0.46	8.95	
	0.00	0.00	0.00	0.00	1.46	24.77	
	0.55	82.88	0.82	96.64	6.64	71.60	
	0.97	232.42	1.92	219.39	8.38	212.88	

backward direction as previously described. The 6-min means are also compiled in their own result file which can be called by the superordinate process control computer.

The whole software of the recording system is concentrated in a real-time program system which was completed by input/output routines of the standard monitor belonging to the system. Considering the presented recording system in connection with the process control computer PRS 4000 and the satellite computer, the whole complex of the integrated data acquisition and primary processing may be regarded a system of measured value acquisition with distributed intelligence.

The described automatic registrating system works continuously since September 1983. It records the 6-min mean values of the geomagnetic components X and Y and, because of the very small magnitude of the short periodic variations in horizontal direction, those of the earth current in east-west direction. Table II shows examples of the output of the system. In the first, third and fifth column of the table there are printed the amplitude mean values in nT, in the second, fourth, and sixth the period mean values in sec, by the way, sequence of components is X, Y, EY. The first line of every mean value group shows the mean values of the pc 2 band, the second those of the pc 3 band and so on. In the ranges pc 2 to pc 4 mean values are calculated every 6-min, because of the possible periods mean value calculation for the pc 5 band is only carried out every 12-min. In the first line of every group there is also printed the beginning of the interval of mean value calculation.

With the beginning of 1984 the monthly report of the Adolf Schmidt Observatory includes a table, which shows a still more condensated representation of the outputs of the described system. For selected days of every month activity indices for hourly mean values calculated from the system output are computed and published.

References

Auster, V.: 1967, Ein Gerät zur Registrierung des zeitlichen Gradienten des geomagnetischen Feldes', *Erdmagnetisches Jahrbuch Niemegk* 1965, S. 157–173.

Auster, V. and Linthe, H.-J.: 1979, 'Programmsystem zur Primärverarbeitung von Digitalregistrierungen kurzperiodischer geomagnetischer Variationen im Echtzeitbetrieb', *Jahrbuch 1978 des Adolf-Schmidt-Observatoriums für Erdmagnetismus in Niemegk, S.* 133–138.

Lenners, D. and Linthe, H.-J.: 1981, 'Erfahrungen beim Echtzeitbetrieb zur Erfassung kurzperiodischer Variationen', *Veröffentlichungen des Zentralinstituts für Physik der Erde Nr. 70*, Teil 2, Potsdam, S. 434–437.

Linthe, H.-J.: Messwerterfassung geomagnetischer Grössen – Ein Beitrag zur Theorie und Praxis rechnergestützter Messtechnik', (not published)

Monthly report of the Adolf Schmidt Observatory for geomagnetism in Niemegk, January 1984.

THE RECORDING OF Pc1 GEOMAGNETIC PULSATIONS USING A MICROCOMPUTER PREPROCESSING SYSTEM

B. J. FRASER and P. W. McNABB
Physics Department, Newcastle University, N.S.W. 2308, Australia

Abstract. In the past slow-speed frequency-modulated analogue tape recording techniques have provided the most convenient method for recording ground-based Pc1 pulsation data on a continuous basis. With events occurring often only a few hours in a week, and a Nyquist frequency of 4–5 Hz, direct digital recording is not practical because of the bulk of data accumulated. However, with the recent availability of reasonably priced microcomputers and advanced signal processing techniques it is now possible to preprocess digital data in the field and store only events of interest. A two-component induction coil magnetometer incorporating a Z80 based 64K RAM microcomputer-floppy disk preprocessing system is described. It is capable of recording Pc1 signals in the 0.2–4 Hz band at middle-to-low latitudes with a minimum detectable signal level of 3 pT. The reliability and limitations of the preprocessing techniques utilizing FFT autospectral analysis to recognize Pc1 signals are discussed.

1. Introduction

Pc1 geomagnetic pulsations are small fluctuations in the earth's magnetic field in the 0.2–5 Hz frequency band. They are discrete emissions generated near the plasmapause region of the magnetosphere by the proton cyclotron instability (see for example, reviews by Gendrin, 1975; 1981). Left-hand mode ion cyclotron waves generated by the instability propagate in wave packets along the magnetic field line direction down to ionospheric heights where some of the energy is transmitted into the ionosphere and the remainder is reflected back along the field aligned path. Left-hand waves are detected at the surface of the earth directly below the ionospheric energy input region (Arnoldy *et al*, 1979). Observations also show that right-hand and linearly polarized waves are seen at higher and lower latitudes (Summers and Fraser, 1972; Hayashi *et al*., 1981). Wave energy propagates to these latitudes in an ionospheric waveguide parallel to the earth's surface and centred on the F2 region (Manchester, 1966).

The dynamic spectra of Pc1 signals often show structured events consisting of a series of overlapping fine-structure bands exhibiting positive or negative slope and time spacings of 0.5–4 min. These structured events tend to be observed more frequently at middle and low latitude than at high latitudes (Kawamura *et al*., 1982). The fine-structure time spacing represents the double-hop transit time for the ion cyclotron wave packet bouncing between hemispheres along the field path. In contrast to this unstructured events are more often observed at high latitudes.

The ionospheric waveguide supports the propagation of Pc1 wave energy over wide areas of the earth's surface. During the International Magnetospheric Study (1976–1979) and more recently a number of spaced station arrays have been established to record Pc1 events propagating between the magnetosphere and the ground (Gendrin *et al*., 1978), between conjugate regions (Glangeaud *et al*., 1980) and over large areas of

Geophysical Surveys **6** (1984) 249-258. 0046-5763/84/0063-0249$01.50.

the earth's surface (Althouse and Davis, 1978; Fraser, 1979; Baransky *et al.*, 1981; Hayashi *et al.*, 1981).

Although the long term sunspot cycle variation in Pc1 wave activity is known (Fraser-Smith, 1981) with the greatest activity being observed between sunspot maxima and minima, precise short term predictions of Pc1 event occurrence are not possible. All that can be stated is that Pc1 events have durations ranging from 10 min to several hours and tend to occur most frequently in the 4–10 day interval following the commencement of major magnetospheric substorm activity (Campbell and Stiltner, 1965). In order to assemble a complete data set of Pc1 event activity it is therefore necessary to record data continuously. The majority of ground-based studies have accomplished this using amplitude-modulated or frequency-modulated slow-speed analogue magnetic tape recording techniques. Tape segments containing Pc1 events of interest were later identified and replayed in the laboratory at audio-frequency tape speeds and analysed using filters and other analogue techniques. Where facilities permitted data were digitized for subsequent time series analysis using spectral and correlation techniques.

Advancement of the experimental knowledge on Pc1 wave generation mechanisms in the magnetosphere and propagation modes in the magnetosphere and the ionosphere over the last few years has been accomplished using recording instrumentation with the ability to record signal arrival times to an accuracy of at least ± 0.05 s. This has been supported by analysis instrumentation and procedures capable of providing high frequency and time resolutions. Modern analysis techniques are necessarily digital. However, with the high data rate associated with a Nyquist frequency of 4–5 Hz, continuous and unattended digital acquisition of Pc1 pulsation data over any significant interval of time is virtually impossible because of the sheer bulk of data accumulated. The alternative method of digitizing recorded analogue data is time consuming as it involves further laboratory processing which generally decreases the quality of the data.

An obvious solution to the data accumulation problem is to either compress or edit the data during the field recording phase. With the recent availability of reasonably priced microcomputers and the real-time implementation of signal processing techniques this is now possible. Compression of multicomponent data into a one dimensional array can be carried out using recently developed polarization state vector techniques (Samson and Olson, 1980; Samson, 1983). As an alternative, editing involves the selection and recording of only the Pc1 events of interest using spectral and/or polarization analysis techniques. The present paper outlines a microcomputer recording system utilizing the data editing technique with the spectral analysis option.

2. The Recording System

The prototype digital recording system is at present operational at the University of Newcastle's field station 30 km North-West of the University campus. Further systems will be installed at the Woomera and Launceston field stations in 1984 and at the remaining stations in our Australian-Antarctic network (Fraser, 1979) in 1985. A block diagram of the complete two-component induction coil magnetometer and digital data acquisition system is shown in Figure 1. Data from both the X (North-South geomagnetic) and Y (East-West geomagnetic) horizontal field components are digitized in the analogue-to-digital converter (ADC) at a rate of $8\,s^{-1}$ providing a Nyquist frequency of 4 Hz. 32 s blocks of X and Y component data are each stored in a 256 byte memory buffer. The data is then preprocessed by a Ferguson Big Board Z80 based microcomputer operating under CP/M with 64K RAM and stored on two double-density 8 inch floppy discs (Figure 2). All timing is controlled by a crystal chronometer accurate to 10 ms day^{-1}. This accurate timing is required in order to measure the extremely small time delay in Pc1 signals arriving at spaced stations, which is of the order of 0–2 s (Fraser 1976).

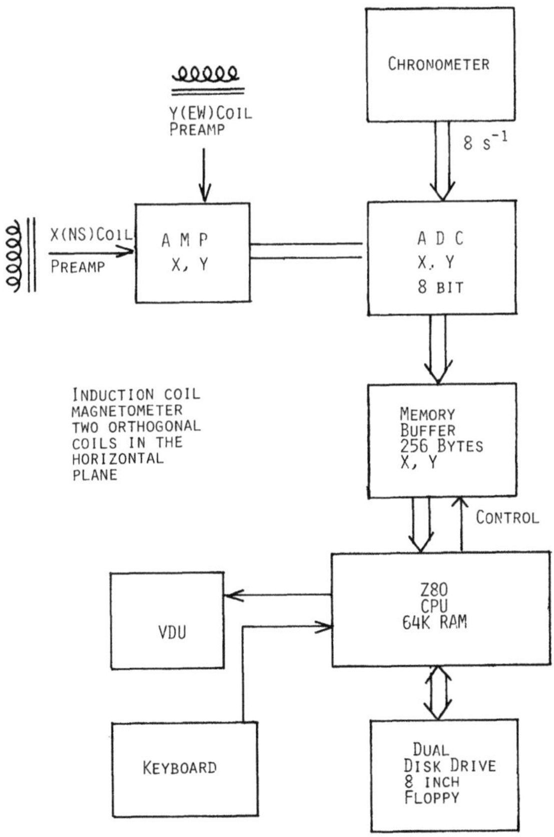

Fig. 1. Block diagram of the two component induction coil magnetometer incorporating a Ferguson Big Board Z80-based preprocessing system.

Fig. 2. The Ferguson Big Board microcomputer. From left to right are the video display unit (VDU), the monitor oscilloscope for the display of digital spectra and the computer and 8 inch floppy disc drive unit.

3. Data Preprocessing Techniques

Pc1 pulsations are band-limited signals of a few minutes to a few hours duration with amplitudes from a few pico-Tesla to a few hundred pico-Tesla at middle and low latitudes. As mentioned earlier, their frequency of occurrence is related to magnetospheric substorm activity but is generally unpredictable, ranging from zero for a week or two up to 10–20 hr on other weeks. In order to digitally record only Pc1 events a real-time data selection protocol must be adopted. In the present application this involves calculating the fast Fourier transform (FFT) power spectrum over 32 s data blocks for the X and Y components using the algorithm of Cooley *et al.*, (1969). The X component power spectrum is then scanned for identification of significant signal peaks above a level set by threshold criteria. At middle-to-low latitudes Pc1 power is seen predominantly in the X component (Summers and Fraser, 1972) and inclusion of the Y component to give the total horizontal field magnitude would only reduce the signal to noise ratio. Recording site external background noise depends on many factors and the threshold level (TL) must be determined at individual stations before setting the data selection criteria. Once this noise level has been established one of two options may be selected for storing data onto disc files. In the first option an arbitrary fixed detection threshold level is set. All 32 s data blocks with one or more spectral estimate in the 0.25–3.75 Hz band exceeding the predetermined level TL are stored on disc. The second option uses a preselected level which is N times the standard deviation of the spectral estimates over the band. This standard deviation will vary from block to block depending on signal power. The parameter N is not necessarily integer and generally in the range 4–7. Again data blocks with power levels above TL are stored. This option has the capability of adjusting TL to a slowly varying background noise level.

There are other Pc1 pulsation properties and data recording characteristics which must be taken into consideration. Pc1 wave amplitude builds up gradually following the initiation of the wave-particle interaction in the magnetosphere and decreases slowly following cessation of the interaction (Fraser-Smith, 1976). There may also be times within an event where the signal drops out for a few minutes or abruptly changes frequency. Transient and impulsive external interference exceeding the threshold level may also occur at any time.

In order to be assured of recording complete Pc1 events and minimal noise the preprocessing system has been designed with two further preset but selectable parameters to take the abovementioned constraints into account. One is the number of 32 s threshold blocks (TB) which must have peak power exceeding TL before a disc file is opened. This precludes the recording of short duration transient and impulsive noise. The other is the number of lead-lag blocks (LB) which must be recorded preceding and following the TB selection. This takes into account the gradual growth and decay of Pc1 event amplitude with time. Data selection was implemented using FORTRAN programming and a schematic example of the technique is illustrated in Figure 3. Shown here is a data block time sequence with four specific situations labelled A to D over 2, 3, 3 and 5 data blocks respectively where the peak power exceeds TL. The preset parameters chosen here were TB = 3 and LB = 4. This requires the peak power to exceed TL for three consecutive blocks before a disc file can be opened and four extra blocks will be recorded before and after these three blocks. In A the disc file does not open since the power in only 2 blocks exceeds TL (TB < 3). The file opens in B since TB = 3 and has a total length of 11 blocks including 4 lead and 4 lag blocks (LB = 4).

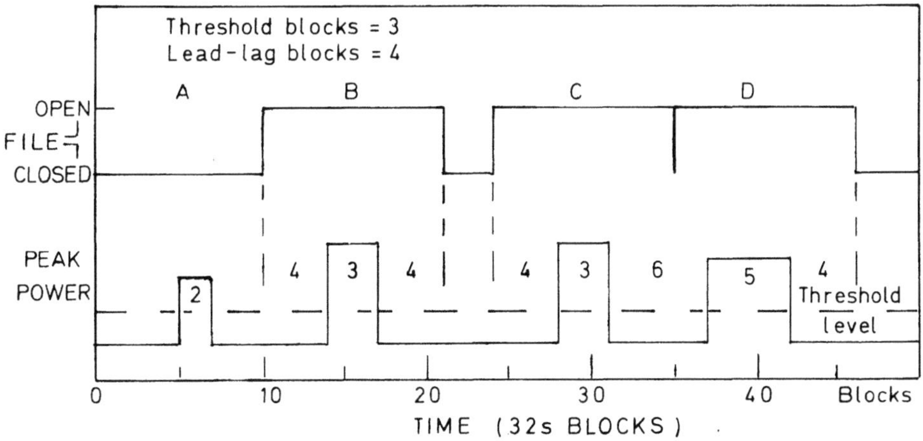

Fig. 3. A schematic diagram of the Pc1 signal detection threshold procedure. The lower plot shows peak power in 32 s data blocks relative to the threshold level (TL). The upper plot indicates the disc file status for four typical conditions A through D. For this example TB = 3 and LB = 4.

The same situation occurs in C with the file closing at block 35. However in D the power exceeds TL after only two further blocks and so a new file is opened at block 35 and closes on block 46. It can therefore be seen that uninterrupted data will be recorded in adjacent files providing the peak power level does not drop below TL for more than 2* LB data blocks.

4. Test Signals and Data

In a data selection system such as that described the minimum detectable signal level will necessarily be TL. It is therefore important to keep this level as low as possible. Three days of continuous data were recorded in order to determine the Newcastle (NC)

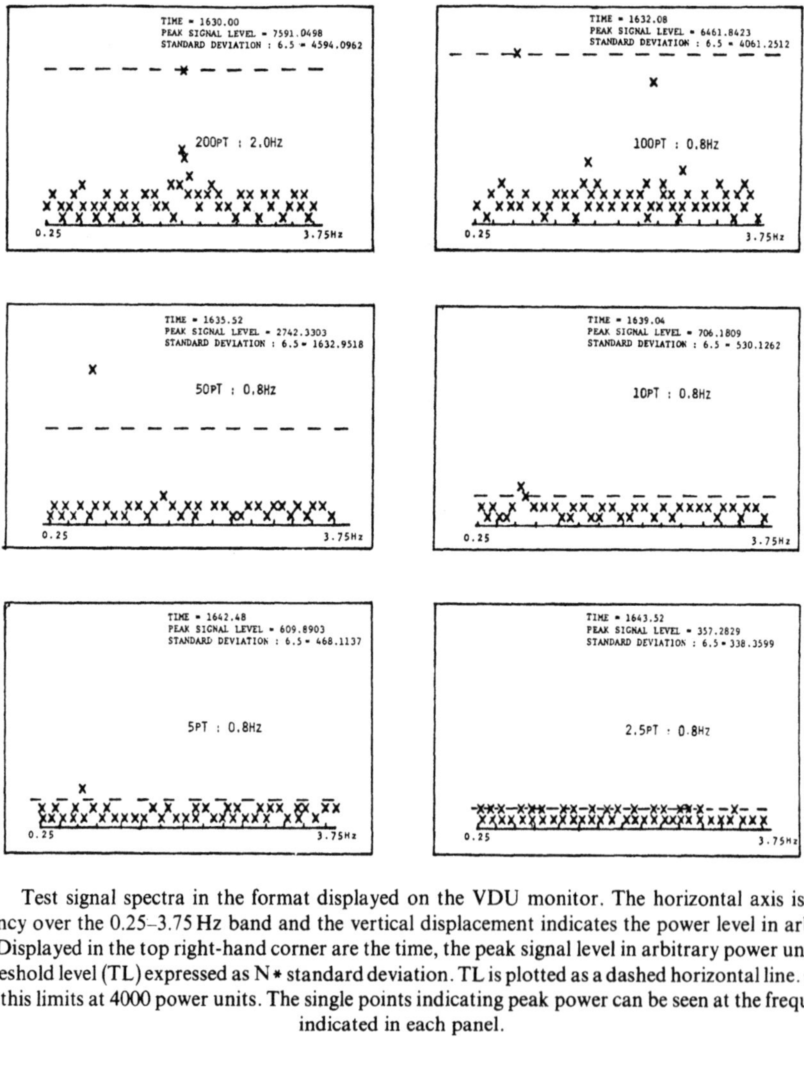

Fig. 4. Test signal spectra in the format displayed on the VDU monitor. The horizontal axis is signal frequency over the 0.25–3.75 Hz band and the vertical displacement indicates the power level in arbitrary units. Displayed in the top right-hand corner are the time, the peak signal level in arbitrary power units and the threshold level (TL) expressed as N * standard deviation. TL is plotted as a dashed horizontal line. On the screen this limits at 4000 power units. The single points indicating peak power can be seen at the frequencies indicated in each panel.

background noise level. Test signals from a standard 250 turn 2 m diameter loop placed at a distance of 50 m from the *X* and *Y* coils and along their axes were used to calibrate the system. Figure 4 shows six 32 s data block panels representing microcomputer VDU monitor power spectra from a series of monochromatic test signals decreasing in amplitude over the range 200–2.5 pT. TL is set at 6.5 standard deviations ($N = 6.5$) and is represented by the horizontal dashed line. The TL level is displayed as the 'standard deviation * 6.5' parameter in the top right-hand corner of each panel. Also displayed are the time in hours, minutes, and seconds and the peak power level in arbitrary power units. The signal level varies between blocks and will obviously influence TL which can be seen to vary between panels. The VDU display has a limiting display value of 4000 power units. It can be seen in the lowest amplitude panel that the background noise level averages about 330 power units. Dynamic spectra of the test signals in Figure 4 are shown in the form of grey-scale plots in Figure 5. The minimum detectable signal is 2.5 pT. A Pc1 event recorded on July 27, 1983 between 1919–2007 hr AEST (09.19–10.07 hr UT) in the frequency band 0.4–0.7 Hz is illustrated in Figure 6. Three consecutive disc files are plotted here. Preceding the event is a 1.25 Hz calibration signal file recorded at 18 hr AEST. These calibrations are recorded every two hours. The following local midnight file with a range of calibration frequencies at 0.5, 0.8, and

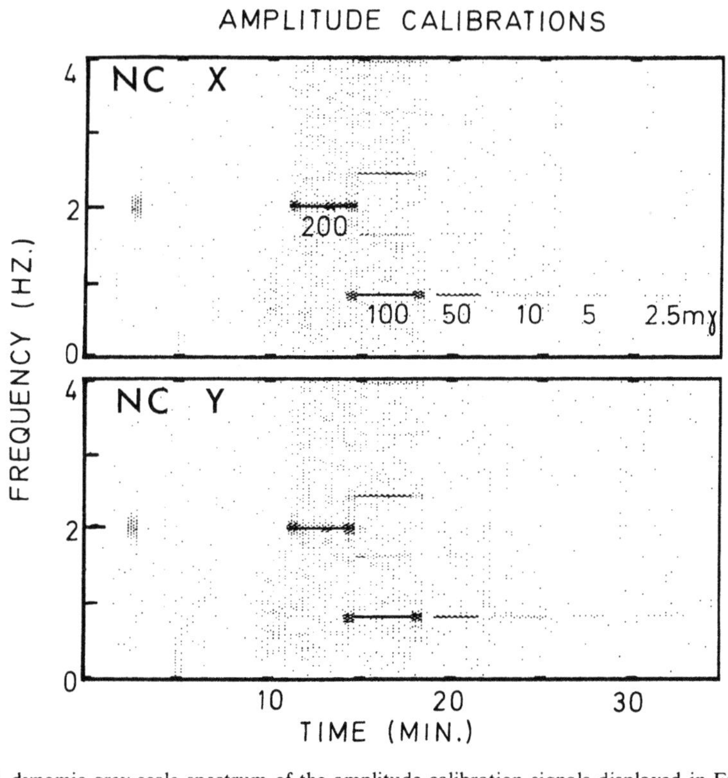

Fig. 5. A dynamic grey-scale spectrum of the amplitude calibration signals displayed in Figure 4. The minimum detectable signal is 2.5 pT.

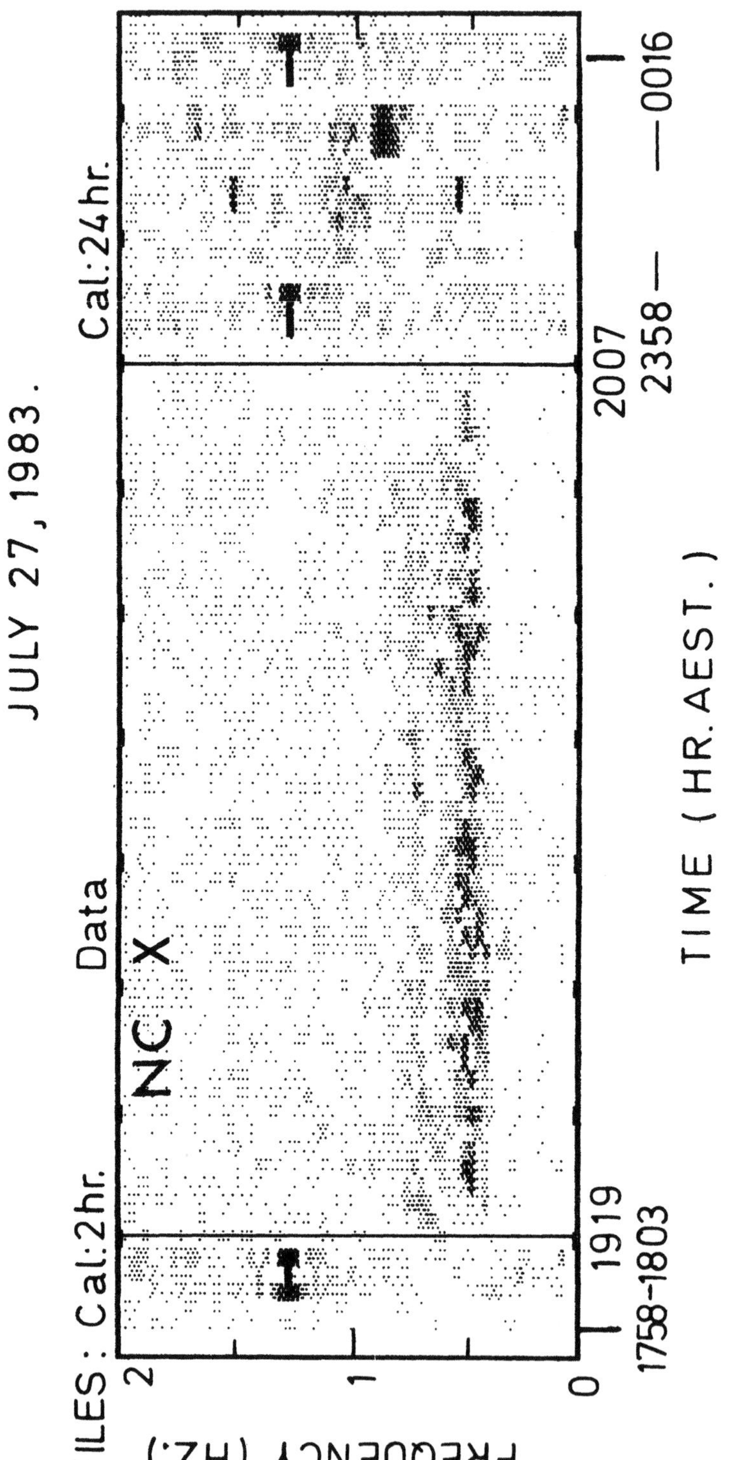

Fig. 6. A dynamic grey-scale spectrum of a Pc1 pulsation event recorded on July 27, 1983. Also shown are the 2 hr and 24 hr calibration files preceding and following the event.

1.25 Hz is also shown. These signals are used for phase corrections in determining the polarization characteristics of Pc1 signals (Fraser *et al.*, 1979). The event in Figure 6 was recorded with the fixed threshold level option set at TL = 350 power units, corresponding to a minimum detectable signal level of 3 pT.

5. Discussion

This preliminary paper has outlined a technique utilizing a standard microcomputer system which may be simply implemented to digitally record Pc1 geomagnetic pulsations. Preprocessing techniques have been employed so that only events of interest are recorded thereby overcoming the data accumulation problem which has limited the digital recording of groundbased Pc1 pulsations in the past. It is important to remember that the system parameters have been designed to record structured Pc1 pulsation events at middle-to-low latitudes and therefore Pc1 and other types of pulsations in the band may not be recorded. Because of this the system would not be suitable for use in synoptic studies.

There are a number of improvements and alternative options planned for the preprocessing system. The data compression techniques inherent in the pure state analysis theory of Samson (1983) can be used to simplify the structure of the data and therefore allow continuous data to be recorded rather than only specific events as described here. At the present time the recording of data is determined using the X component autospectrum for reasons previously explained. The percent polarization parameter obtained from coherency matrix analysis (Fowler *et al.*, 1967) provides a measure of polarized power in a signal and may be a more sensitive parameter for detecting the presence of Pc1 events. These developments will be attempted on the prototype system and an optimum preprocessing system finalized before further systems are deployed at other field stations.

In conclusion, the preprocessing techniques using a microcomputer system described in this paper have shown that Pc1 events can be successfully recorded in digital form with minimal data accumulation. The system would be ideal for use in multistation groundbased networks where only specific events with amplitudes above a predetermined level are required for detailed study.

Acknowledgments

This research was supported by the Australian Research Grants Scheme and University of Newcastle research grants.

References

Althouse, E. L. and Davis, J. R.: 1978, 'Five-Station Observations of Pc1 Micropulsation Propagation', *J. Geophys. Res.* **83**, 132–144.

Arnoldy, R. L., Lewis, P. B., and Cahill, L. J.: 1979, 'Polarization of Pc1 and IPDP Pulsations Correlated with Particle Precipitation', *J. Geophys. Res.* **84**, 7091–7098.

Baransky, L., Golikov, Yu., Feygin, F., Harchenko, I., Kangas, J., and Pikkarainen, T.: 1981, 'Role of the Plasmapause and Ionosphere in the Generation and Propagation of Pearl Pulsations', *J. Atmos. Terr. Phys.* **43**, 875–881.

Campbell, W. H. and Stiltner, E. C.: 1965, 'Some Characteristics of Geomagnetic Pulsations at Frequencies near 1 c/s', *Radio Sci. Res.* **69D**, 1117–1132.

Cooley, J. W., Lewis, P. A. W., and Welch, P. D.: 1969, 'The fast Fourier Transform and its Applications', *IEEE Trans. Ed (U.S.A.)* **12**, 27–34.

Fraser, B. J.: 1976, 'Pc1 Geomagnetic Pulsation Source Regions and Ionospheric Waveguide Propagation', *J. Atmos. Terr. Phys.* **38**, 1141–1146.

Fraser, B. J.: 1979, 'The Use of Multistation Networks in Geomagnetic Pulsation Studies, *Ann. Telecomm.* **34**, 220–236.

Fraser, B. J., Kemp, W., and Webster, D. J.: 1979, 'Ionospheric Duct Propagation and Pc1 Source Regions', *Proc. Int. Workshop on selected Topics of Magnetospheric Phys.*, Tokyo., Japanese IMS committee, pp. 181–183.

Fraser-Smith, A. C.: 1976, 'Measured Growth of Pc1 Geomagnetic Pulsations', *EOS* **57**, 983.

Fraser-Smith, A. C.: 1981, 'Long-term Predictions of Pc1 Geomagnetic Pulsations: Comparisons with Observations', *Planet. Space Sci.* **29**, 715–719.

Gendrin, R.: 1975, 'Waves and Wave-Particle Interactions in the Magnetosphere: A Review', *Space Sci. Rev.* **18**, 145.

Gendrin, R.: 1981, 'General Relationships between Wave Amplification and Particle Diffusion in a Magnetoplasma', *Rev. Geophys. Space Phys.* **19**, 171–184.

Gendrin, R., Perraut, S., Fargetton, H., Glangeaud, F., and Lacoume, J. L.: 1978, 'ULF Waves: Conjugated Ground-Satellite Relationships', *Space Sci. Rev.* **22**, 433–442.

Glangeaud, F., Lacoume, J. L., Fargetton, H., Gendrin, R., Perraut, S., and Troitskaya, V. A.: 1980, 'Cross-Spectral Analysis of Pc1-2 Emissions at Different Stations', *J. Geophys. Res.* **85**, 4115–4124.

Hayashi, K., Kokubun, S., Oguti, T., Tsuruda, K., Machida, S., Kitamura, T., Saka, O., and Watanabe, T.: 1981, 'The extent of Pc1 Source Region in High Latitudes', *Can. J. Phys.* **59**, 1097–1105.

Kawamura, M., Kuwashima, M., and Toya, T.: 1982, 'Comparative Study of Magnetic Pc1 Pulsations Observed at Low and High Latitudes: Source Region and Generation Mechanism of Periodic Hydromagnetic Emissions', *Memoirs Nat. Inst. Polar Res.* **22**, 3–16.

Manchester, R. N.: 1966, 'Propagation of Pc1 Micropulsations from High to Low Latitudes', *J. Geophys. Res.* **71**, 3749–3754.

Samson, J. C.: 1983, 'The Spectral Matrix, Eigen Values and Principal Components in the Analysis of Multichannel Geophysical Data', *Annls. Geophys.* **1**, 115–120.

Samson, J. C. and Olson, J. V.: 1980, 'Some Comments on the Descriptions of Polarization States of Waves', *Geophys. J. Roy Astron Soc.* **61**, 115–129.

Summers, W. R. and Fraser, B. J.: 1972, 'Polarization Properties of Pc1 Micropulsations at Low Latitudes', *Planet. Space Sci.* **20**, 1323–1335.

The second group of papers describes various aspects of geomagnetism in the field. Again range of application is the impressive element of the collection. A land survey in Czechoslovakia up-dating the country's magnetic chart for secular variation, an airborne survey to extend knowledge of the magnetic field to the seas around Japan and a discussion of methods to ensure that such observations are corrected for time variations of the field in the best possible way. Two papers illustrate applications of magnetic observations in one case to support EISCAT, a multi-national radar experiment measuring parameters of the ionosphere and magnetosphere, and the other using natural rapid electromagnetic variations in a geophysical survey of the structure of the Earth's crust (at a vulcano). The collection highlights very clearly the value of a clear understanding of the geomagnetic field and its variations and the ability to make appropriate measurements of it to a very diverse range of scientific interests.

MAGNETIC MAPPING OF SLOVAKIA FOR THE EPOCH 1980.5

J. PODSKLAN

Geophysical Institute, Slovak Academy of Sciences, 94701 Geomagnetic Observatory, Hurbanovo, Czechoslovakia

Abstract. 130 sites were occupied in a magnetic survey of Slovakia. About half of those occupied in previous surveys (1952.5, 1958.5, and 1967.5) were unsuitable because of increased man made noises. New stations were substituted. A combination of classical magnetometers (for *H* and *D*) and a proton magnetometer (for *F*) was used. *Z* was computed from *H* and *F*. Instruments were calibrated at Hurbanovo magnetic observatory before the survey and at regular size intervals during it.

Reduction to epoch was made on the assumption that the scale variation is everywhere the same and equal to that at Hurbanovo. Preliminary maps are hand drawn, but the data have been transferred to computer tape and computer plotted maps are being prepared.

1. Introduction

The first magnetic measurements were made in Slovakia (see Figure 1.) as part of the magnetic mapping of Hungary in the years 1843–1848. In these measurements, and in others carried out in 1864–1879, 1892–1894, 1928–1929, and 1936–1937, the number of stations was not large and the organization of the observations was not designed to determine all the components of the Earth's magnetic field.

The first mapping of the 1st order was carried out by Š. Ochaba [1] in the years 1951–1953 who reduced it to the epoch 1952.5, and the second in the year 1958 which he reduced to the epoch 1958.5 [2].

In the years 1967 and 1968, a team from the Geophysical Institute of the Slovak Academy of Sciences carried out measurements at 132 points distributed throughout Slovakia. These measurements were reduced to the epoch 1967.5 [3].

The mapping which was carried out in the years 1980–1982 and reduced to the epoch 1980.5 ties in with the results of the projects mentioned above.

The mapping consisted of three main stages:

(1) Preparatory stage.
(2) Field measurements.
(3) Processing of results.

2. Preparatory Stage of Mapping

The preparatory stage consisted of checking the stations at which measurements were made during the previous mapping in 1967 and 1968 as regards suitability for the contemplated magnetic measurements and of seeking new stations in place of the old ones which proved unsatisfactory. This stage was carried out in 1979.

From the magnetic point of view, each station was checked by measuring with a proton magnetometer directly above the station mark of the station and at four points

Geophysical Surveys **6** (1984) 261-267. 0046-5763/84/0063-0261$01.05.

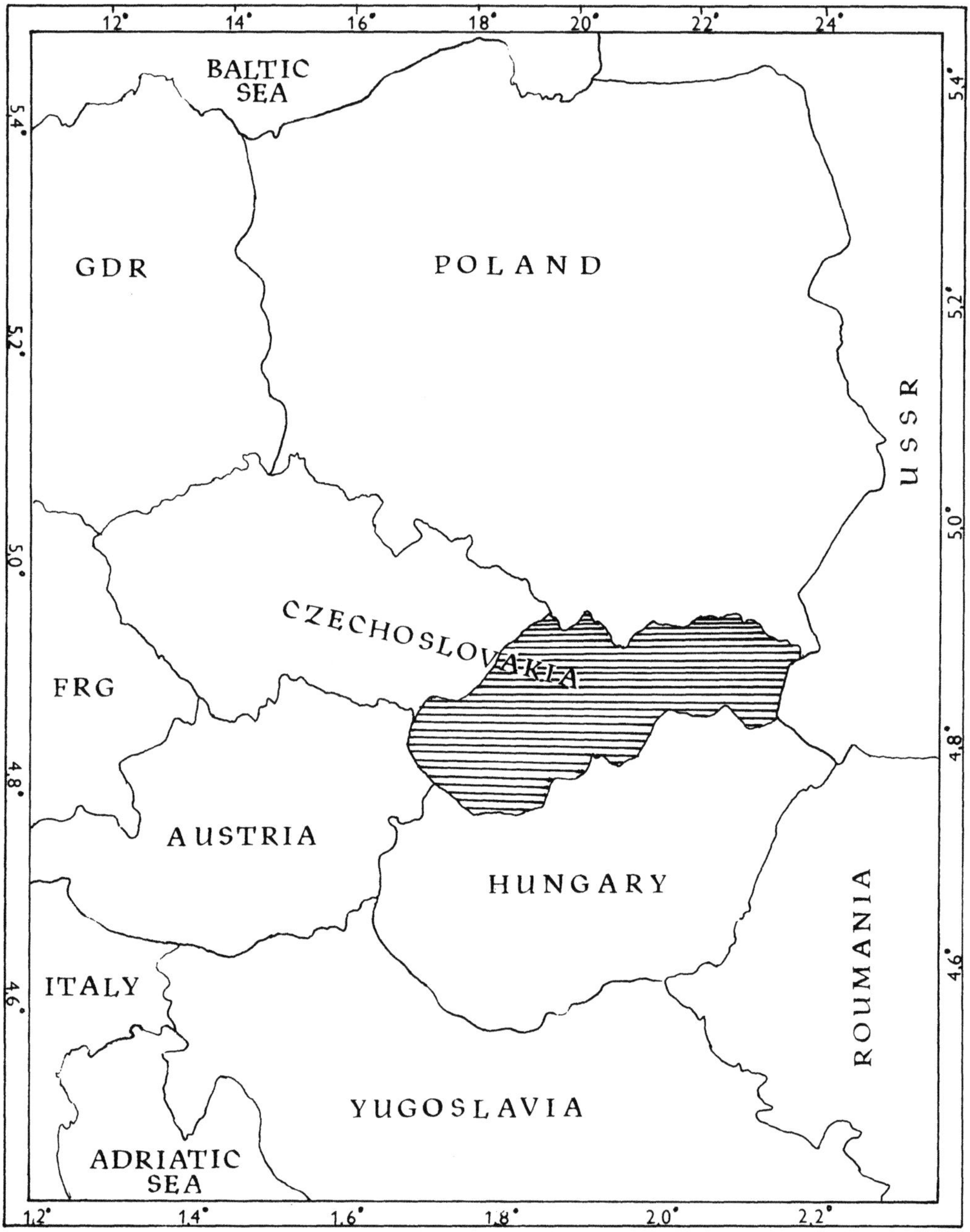

Fig. 1. The territory of Slovakia is indicated by horizontal shading.

roughly 5 m distant from the station in the directions of the principal points of the compass, the purpose being to determine whether the field was homogeneous in the station's immediate vicinity. The check measurements were reduced to the epoch 1979.5 and compared with the results of previous measurements reduced to the epochs 1952.5 and 1967.5.

A great advantage in selecting the stations was the design of the ELSEC 592 proton magnetometer used, which enables the occurrence of man-made disturbances to be detected. This allowed suitable time intervals to be chosen for measurements even in areas with a high level of industrial noise.

In the course of the preparatory work it was found that of the original 132 stations only 52, i.e. 39.4%, were suitable for further measurements, 8.3% being partly disturbed and 52.3% of the stations being quite unsuitable.

The unsuitable stations were substituted by new ones. A magnetic network was thus created, consisting of 130 points throughout Slovakia. This represents one point per 370 km^2.

The large number of stations which proved unsuitable for further magnetic measurements, indicates what problems would be encountered in long-term detailed measurements conducted for the purpose of determining anomalies of the secular variation on the territory in question.

3. Instrumental Equipment

We considered, first of all, the choice of instruments. We tested two possibilities: measurements with a vector proton magnetometer and measurements with classical instruments, the total field vector being determined by proton magnetometer. We decided in favour of the latter.

The measurement of components with a proton magnetometer requires the use of compensation coils. When used for observatory purposes, no particular attention need be given to their size. However, in field measurements, it is important that the coils be sufficiently small and sufficiently stable to guarantee a sufficiently homogeneous field. A further problem is encountered in stabilizing them in the field. The error incurred due to the tilt of the coil axes when the observer moves represented about 2 nT and could not be eliminated by the measuring procedure. Moreover, it has a random character. Another unfavourable factor was the air flow causing the coil axes to tilt. Under field conditions in Central Europe it is difficult to make observations only at times when there is no wind. Observations at each new station require repeated setting of the magnetic axis of the coils relative to the magnetic meridian. This increases the time required for one measurement considerably.

Some of these disadvantages could be overcome partly or completely. For example, the observation stations could be arranged to eliminate the effect of the observer moving round the instrument. However, in the case of a large number of stations this would prove very exacting for a number of reasons.

Having considered these facts, we decided in favour of the second possibility, i.e. of

TABLE I

Instruments used in geomagnetic mapping of Slovakia and their assumed accuracy

Observed quantity	Variations	Azimuth	Declination	H-component	Total field F	Z-component
Instrument	Bobrov field set D, H, Z	Zeiss Theo 010	Field theo Mating and Wiesenberg	3 QHM's	Proton magnetometer ELSEC 592	ELSEC + QHM's
Accuracy	–	$\pm 0.1'$	$\pm 1'$	± 3 nT	± 0.5 nT	± 2 nT

combining classical instruments to measure the components with a proton magnetometer to measure the total field. Consequently, we used the following equipment for the magnetic measurements: 3 QHM's for measuring the horizontal component, 1 Mating and Wiesenberg field theodolite for measuring the declination and an ELSEC 592 proton magnetometer for measuring F (see Table 1). A Zeiss Theo 010 theodolite was used for observing the Sun in determining the astronomical azimuth. Time was measured with a stop watch and a half-second chronometer.

4. Accuracy of Observations

To be able to estimate the accuracy of observations made by classical instruments, we laid out a polygon, consisting of three points, in the precinct of the Geomagnetic Observatory in Hurbanovo. We made several tens of observations at the points of this polygon and reduced them with respect to variation observations. We arrived at the following errors of observations of the separate components (see Table I): $\pm 1'$ for D, ± 3 nT for H and ± 2 nT for Z.

The proton magnetometer is compared regularly every two years at the Adolf Schmidt Geomagnetic Observatory in Niemegk. The estimated accuracy is ± 0.5 nT. Also the classical instruments are checked at this observatory every two years. As regards the mapping, this means comparative measurements in 1980 before beginning the mapping, and in 1982 on terminating it. During the mapping, the classical instruments were also checked before taking to the field and after returning from field measurements by comparing with the Hurbanovo observatory standard. Since the individual field sorties amounted to about 7 days, this represented one check every week. This guaranteed the accuracy with which the instruments measured.

5. Measurements

Absolute measurements were made at the same height above the terrain at all stations.

Observations of the total field consisted of observations at the station itself (10 readings of F) and the probe was then quickly moved to an auxiliary point, about 50 m distant, where the total field was measured again. This determined the difference between the auxiliary and main station. Moreover, the total field was measured at the auxiliary station at the times when readings with the QHM instruments were being taken. This provided data on the variation of the total field at the point of observation over an interval of roughly 30 min.

Observations with the QHM's were made in the usual way: each of the instruments was used for one observation series. Also observations of the declination were carried out in the usual way, using the field theodolite.

The vertical component was measured indirectly by reading the total field at the auxiliary station at the time of the QHM reading. The Z-component was then calculated, using the well-known formula, from the H-component and F, after eliminating the correction of F for the auxiliary station. This method of determining Z has the advantage that it requires no additional instrument. Practically no time is required for the measurement either, and the accuracy is relatively high.

The azimuth of the azimuth mark was determined by observing the Sun. Three series of measurements were made at each station. With a very few exceptions, these observations were carried out outside of the interval of 2 hr either side of local noon.

In reducing the observations, the time variations of the Hurbanovo Observatory variometers were used for most stations. Observations from a smaller number of stations were reduced with the aid of field variation located in situ.

Magnetograms from field variation stations were used in the case of 29 points, i.e. 22.5% of all points. The differences of the values reduced on the basis of observatory magnetograms with respect to those reduced using the field stations for declination are in the interval $-1.0'$ to $+1.1'$.

The mean square deviations for all 29 points, comparing the reductions using the field stations and those using the observatory values, are $\sigma_D = 0.55'$ for D, $\sigma_H = 1.52$ nT for the H-component and $\sigma_Z = 0.7$ nT for the Z-component.

6. Processing of Observations

The observed values were used to set up data sets which were stored on magnetic tape. For further processing the data on the tapes were loaded into the computer memory and, if required, supplemented with calculated values.

The observations were reduced to the epoch 1980.5 under the assumption that the secular variation of the Earth's magnetic field was the same throughout Slovakia and equal to the secular variation at Hurbanovo. We used the formula

$$E^{\text{point}}_{1980.5} = E^{\text{point}}_{t} - \left(E^{\text{HRB}}_{t} - {}^{\text{HRB}}_{1980.5}\right),$$

where E is the appropriate element, t is time at which the reading was taken and HRB is the reference point Hurbanovo.

The results of the mapping of Slovakia are maps. The preliminary maps were drawn by hand. The observed and reduced data will be finally processed by computer. The computer output will be connected to a plotter which will directly plot complete

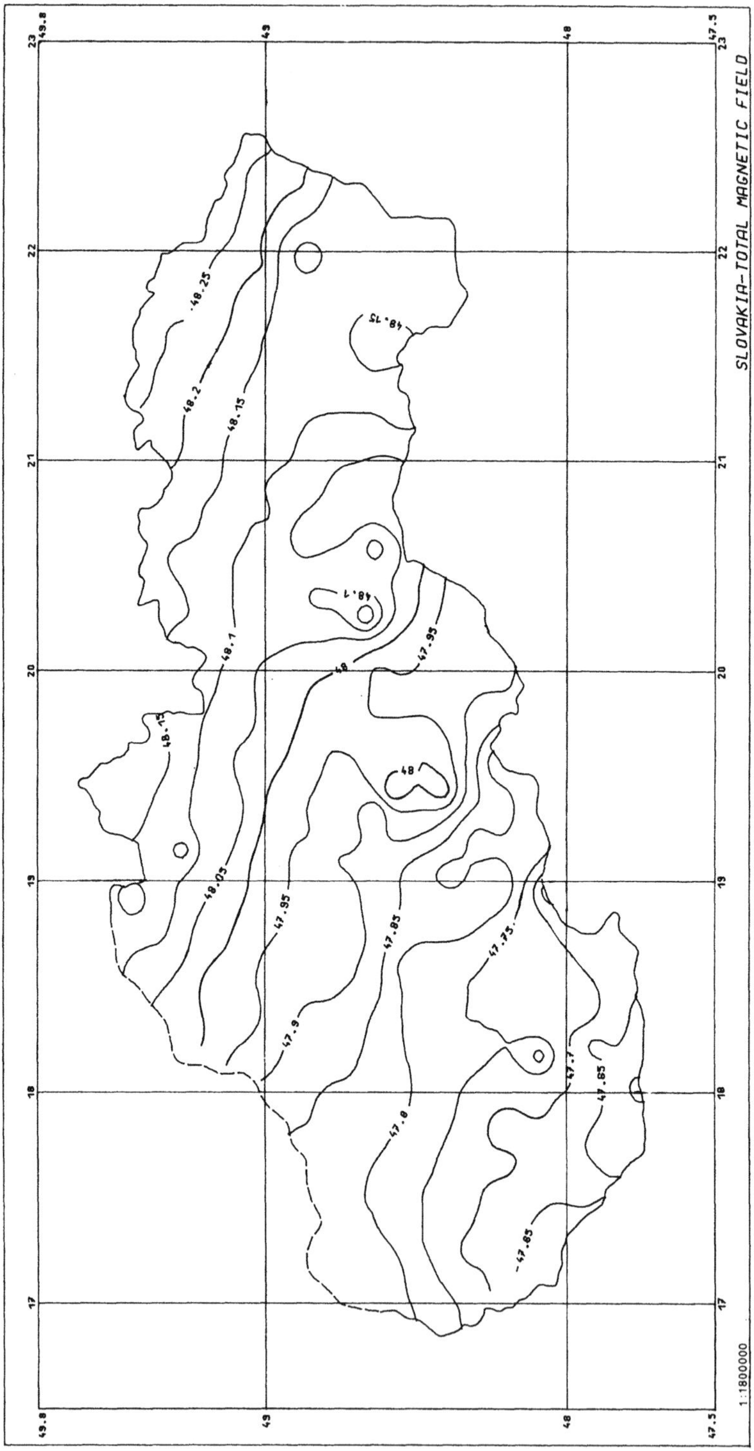

Fig. 2. The preliminary map of the total field F on the territory of Slovakia. The values are given in μT ($1\ \mu T = 10^{-6}\ T$).

magnetic maps. In calculating the isolines, the iteration method will be used. The preliminary map of the total field *F* generated by computer is given in Figure 2.

7. Conclusion

The purpose of this paper was to draw attention to some of the problems which now occur in magnetic mapping: increasingly more serious problems with seeking undisturbed stations for observation; there are still no instruments for field observations which would enable as simple and accurate observations of the components as of the total field *F*.

The maps of the individual elements will be used for comparison with the distribution of other geophysical fields on the territory of Slovakia.

References

[1] Ochaba, S.: 1958, 'Distribution of Geomagnetic Field in Slovakia for the Epoch 1952.5', *Travaux Géophysiques* No. 72–98, pp. 319–362, Praha 1959.

[2] Ochaba, S.: 1979, 'Contribution to the Study of the Geomagnetic Field and Its Secular Variation Distribution in Slovakia', *Acta fac. rerum. natur. Univ. comen. Astron. et geophys.* **5**, 25–39.

[3] Krajčovič, S. and Németh, M.: 1972, 'Distribution of the Geomagnetic Field in Slovakia for the Epoch 1967.5', *Contributions of the Geophys. Inst. of the Slovak Academy of Sciences* **3**, 16–24.

GEOMAGNETIC ABSOLUTE MEASUREMENTS WITH A NUCLEAR RESONANCE THEODOLITE AT THE ADOLF SCHMIDT OBSERVATORY IN NIEMEGK

V. AUSTER

Central Institute of Solar-Terrestrial Physics, Acad. Sci. G.D.R., G.D.R.-1199, Berlin

The instrumentation for getting digital minute values of the components F, Z, and Y of geomagnetic variations which has been successfully operated at the Adolf Schmidt Observatory for some years was, in 1982, completed by an absolute device based on the nuclear resonance effect. This device measures the total intensity F, the vertical component Z, the horizontal intensity H and the component Y directed geographically east. For the generation of auxiliary fields a special cylindrical coil rotatable around vertical and horizontal axes is used. The coil axis is always situated in the horizontal plane. For the magnitude of the auxiliary field a fixed value is used which is chosen to be approximately 2H for technical reasons. The mechanical part of the device is accomodated in the house for absolute measurements. A keyboard for remote control, connected with the electronic part of the instrument, is installed in another building at a distance of approximately 80 m. The measurement results are recorded along with the actual time on a special punched tape. The resolution is 0.1 nT. Additionally, a special punched tape containing output from the variometer station with the same resolution can be produced.

Measurements are arranged in such a way that one measuring series lasts for a maximum time of one minute. This applies to the measurement of components, with the auxiliary field reversed between adjacent measurements (8 measurements). For measuring F. (4 measurements), half a minute is sufficient. The other half-minute is used for adjusting the device in the north-south direction and in the direction to the fixed mark for measuring Y. The whole cycle for a complete measurement covers 2 measurements in each case for H (or Z) where in each case the measurements are carried out in the NS and SN directions and analogously 2 measurements for Y (in each case in the EW and WE directions). Thus, 17 min are necessary for one complete measurement. The values for H, Z, Y, and the mean values for F and the measured value for the auxiliary field (for checking) are, together with an indication of the variations from the punched tape, directly computed by our PRS 4000 computer using a special program and are printed.

The reduction of the variation values on the punched tape produced during the measuring time is then also carried out by a computer program. The recorded values, some important intermediate results and, finally, the desired absolute values for F, Z, H, and Y (on a special computer record) are also printed out. It turns out that it is necessary to check the levelling of the device before each measurement and to correct it

Geophysical Surveys **6** (1984) 269-270. 0046-5763/84/0063-0269$00.30.

if necessary. The vertical position of the standing axis of the theodolite and the exact horizontal position of the mechanical axis of the auxiliary field coil are checked at greater temporal intervals as well as the parallelism of the magnetic coil axis to the telescope axis.

Since the measuring devices for the standard and nuclear resonance measurements of absolute values are installed in the same building, simultaneous measurements are not possible. Comparisons between both measuring devices are therefore carried out by means of geomagnetic normal recordings for the variation. The following are the first preliminary values for March 1983 (compiled from 6 measurements):

$$F = -0.8 \pm 0.6\,\text{nT};$$
$$Z = 2.2 \pm 0.5\,\text{nT};$$
$$H = 3.4 \pm 0.6\,\text{nT};$$
$$Y = 12.3 \pm 1.7\,\text{nT}.$$

At present the quality of the Y measurement is still being improved. A routine application of nuclear resonance absolute measurements in parallel with the standard method will start in 1984.

AEROMAGNETIC SURVEYS IN THE SEAS AROUND JAPAN IN 1980

S. UTASHIRO
Department of Physics, Science University of Tokyo, Tokyo, Japan

S. OSHIMA and T. KANEKO
Hydrographic Department, M.S.A., Tokyo, Japan

Abstract. Aeromagnetic surveys in the seas around Japan were carried out in 1979 to 1980 by a YS–11 aircraft of the Maritime Safety Agency of Japan using a new type of airborne magnetometer system. The new type of airborne magnetometer system consists of a ring-core type three-component fluxgate magnetometer, an inertial sensing system, a fish-eye camera to measure the true north, an 8-bit microcomputer and a proton magnetometer. The VLF/OMEGA system is used to fix the position of the aircraft. Tracks extended to about 600 nautical miles off the coast of the Japanese islands in the Sea of Japan and the North-West Pacific Ocean. Average spacing between tracks was about 80 miles. The flights were carried out at an altitude of 9500 feet. From the survey results, magnetic charts of the seven geomagnetic elements for 1980 over the sea around Japan were compiled by the method of least squares using a polynomial. Also, the contour charts of secular variation in 1980 were compiled.

1. Introduction

The primary object of a geomagnetic survey is to acquire accurate data on the Earth's magnetic field. The data are useful for navigation and can be employed for scientific investigation. In order to determine the variation of the magnetic compass, the Hydrographic Department of Japan has performed land magnetic surveys in Japan every 5 yr since 1950. Recently, the aeromagnetic survey technique has been developed in Japan. The Hydrographic Department of Japan[1] [2] [3] has carried out aeromagnetic surveys in the seas around Japan four times since 1962. The first of these aeromagnetic surveys was carried out from 1962 to 1964 by an airborne magnetometer installed in a Beachcraft 502 aircraft as part of the program of the World Magnetic Survey. The airborne magnetometer consisted of a three-component saturable core, gimbals, a servo driving system to keep the detector system vertical and a fish-eye camera to determine true north. The second aeromagnetic survey was carried out from 1969 to 1970 using a fluxgate magnetometer, proton magnetometer, vertical gyroscope and fish-eye camera installed in a long distance aircraft, a YS–11, 701. The sensor part of the fluxgate magnetometer and the proton magnetometer were installed in a plastic body which extended one and half meters astern of the tail of the aircraft. In this survey, the tracks extended to about 500 nautical miles off the coast of the Japanese islands. Magnetic charts for 1965.0 and 1970.0 were compiled on a scale of 1 : 2 500 000 for the seven elements of the Earth's magnetic field D, H, I, F, X, Y, and Z using the observed data from these magnetic surveys.

This paper contains the results of the latest aeromagnetic survey and land magnetic

Geophysical Surveys 6 (1984) 271-290. 0046-5763/84/0063-0271$03.00.

survey in Japan carried out in 1979 and 1980. Aeromagnetic surveys in the seas around Japan were carried out by a YS–11 aircraft belonging to the Maritime Safety Agency of Japan using a new type of airborne magnetometer system. The new type of airborne magnetometer system consists of a ring-core type three-component fluxgate magnetometer, a proton magnetometer, an inertial sensing system containing an electrical gyroscope, a fish-eye camera to measure the true north and an 8-bit microcomputer. The VLF/OMEGA system was used to fix the position of the aircraft at sea. The tracks extended to about 600 nautical miles off the coast of the Japanese islands in the Sea of Japan and the North-West Pacific Ocean. The average spacing between tracks was about 80 miles. The flights were carried out at an altitude of 9500 feet. The normal distribution of the seven elements of the Earth's magnetic field, D, H, I, F, X, Y, and Z, were represented by a third-degree polynomial with respect to latitude ϕ and longitude λ using the data observed during the aeromagnetic and land magnetic surveys. The calculations were carried out by a large electronic computer, an ACOS 700. Also, the secular variation in the seas around Japan in 1980 was determined using the data at fourteen magnetic observatories around Japan.

2. Aeromagnetic Survey

2.1. Aircraft used and method of observation

The aircraft used for the survey was a YS–11 which had been modified to increase its range and to make it suitable for magnetic surveying. The aircraft was equipped with a tail-stinger of non-magnetic plastics, 1.5 m long at the rear of the fuselage. As for the general specification of the aircraft, its length was 26 m, wing span 32 m, cruising speed 220 knots and flying range about 1200 nautical miles (when carrying the magnetometer). It was equipped with an auto-pilot system. The instruments used for the survey were a proton magnetometer, a three-component airborne magnetometer and VLF/OMEGA receivers. Any error in position fixing was estimated to be less than two nautical miles even in the worst conditions.

Measurements with the proton magnetometer were made once every two seconds with an accuracy of ± 1 nanotesla. The three-component airborne magnetometer consisted of a fluxgate magnetometer to measure three elements of the magnetic field, an inertial sensing system to measure the roll, pitch and yaw angles of the aircraft, and an azimuth measurement system to take photographs of the position of the sun by a fish-eye camera.

The survey data: position, three components of the magnetic field, attitude and heading of the aircraft were recorded once a second on a magnetic tape cassette. The heading direction was corrected by measurements of true north carried out every minute with the fish-eye camera.

Installation of the equipment on the aircraft was made as follows. The ring-core type sensor of the fluxgate magnetometer was fixed inside the tail-stinger at the rear of the fuselage. The fish-eye camera was attached to the ceiling in the fore part of the fuselage, and the other instruments were installed in the after-cabin. The locations of the flight lines are in Figure 1.

2.2. Determination of instrumental errors

2.2.1. *Magnetometer*

The instrumental errors were determined each year before and after the survey by comparing the instruments with the standard magnetometer KASMMER at the

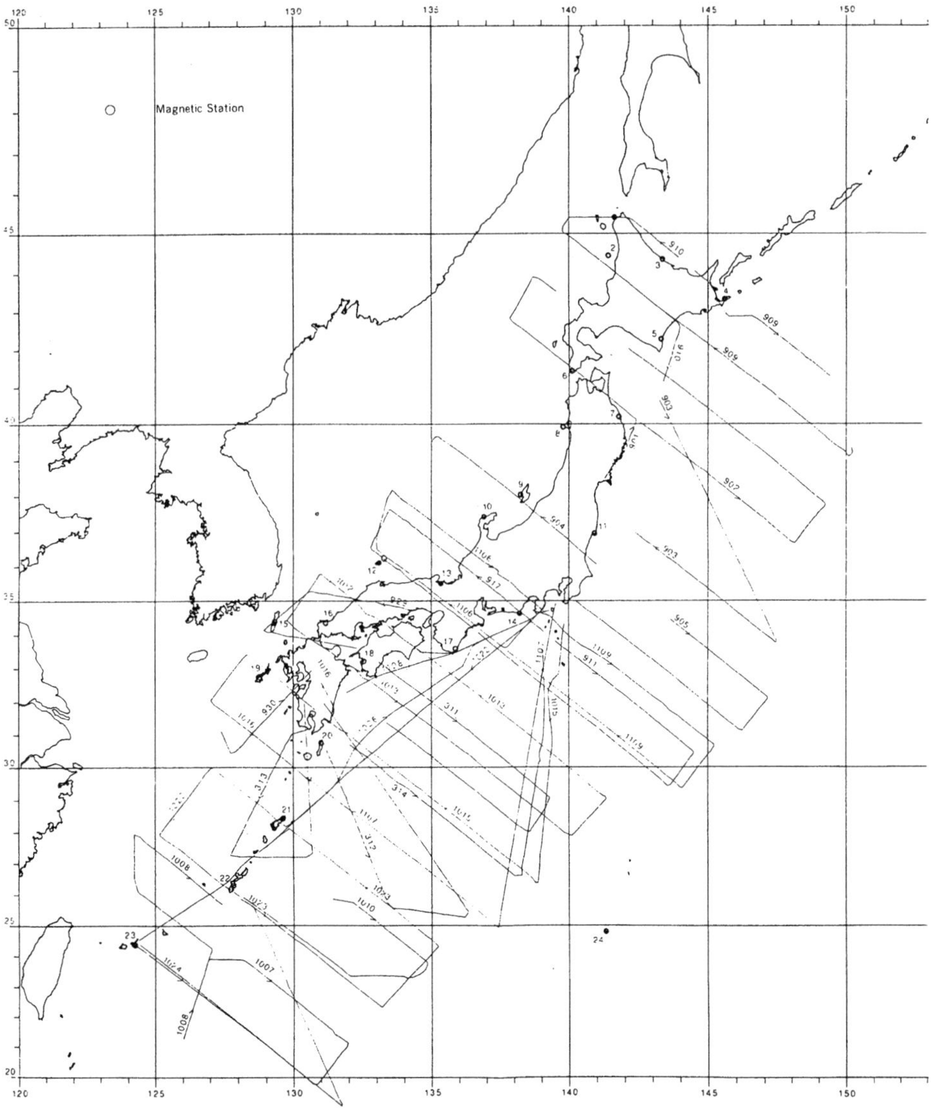

Fig. 1. Distribution of Magnetic Stations and Tracks for the Aeromagnetic Surveys.

Kakioka Magnetic Observatory of the Japan Meteorological Agency. As far as the proton magnetometer was concerned, no instrumental error was detected. In this paper an explanation will be given of the instrumental errors of the fluxgate magnetometer. In the sensor of the fluxgate magnetometer, three detecting coils are fixed to measure the elements of the magnetic field along three directions mutually intersecting at right angles, one forward along the fore-and-aft axis of the aircraft, one to larboard in the port-and-larboard direction and one downward and normal to the other two axes. In the ideal case, the electrical axes of these coils mutually intersect exactly at right angles. Actually, however, the axes were slightly out of the ideal positions, so that corrections for this should be made.

The comparative measurements with the standard magnetometer were conducted in an absolute observation house. The sensor of the fluxgate magnetometer was fixed on a non-magnetic rotating table resembling the horizontal circle of a theodolite. The observations were made on an observation platform. To obtain instrumental errors, two coordinate systems are considered. Firstly, the rectangular coordinate system fixed to the sensor is defined as the *abc* coordinate system where the *a*-axis, when the sensor is installed on the aircraft, corresponds to the direction forward along the fore-and-aft axis of the aircraft, the *b*-axis to larboard in the port-and-larboard direction and the *c*-axis downward and normal to the other two axes. Secondly, the non-rectangular coordinate system with the axes of the three detecting coils not intersecting at right angles to one another is defined as the $a'b'c'$ coordinate system. The two systems are so arranged that axes a and a', b and b', and c and c' are almost parallel to each other, respectively. Let us assume that the components of the magnetic field along the directions of axes a, b, and c are A, B, and C, respectively and that the measured values of the magnetic components along the directions of axes a', b', and c' in the fluxgate magnetometer are A', B' and C', respectively. Then, it is assumed that there are the following relationships between these values:

$$\begin{aligned} P_a \cdot A + Q_a \cdot B + R_a \cdot C &= A' \\ P_b \cdot A + Q_b \cdot B + R_b \cdot C &= B' \\ P_c \cdot A + Q_c \cdot B + R_c \cdot C &= C' \end{aligned} \tag{1}$$

where P_i, Q_i, and R_i ($i = a, b, c$) are the coefficients for correcting for the non-rectangularity of the axes of the detecting coils. The values of A, B, and C may also be obtained from the standard magnetometer and the magnetograms at the observatory. In the comparison measurement, the rotating table on which the sensor was mounted was rotated every 5 degrees of angle, and about 500 sets of measurements were conducted. Each of the correction coefficients was obtained by the method of least squares. Results of the measurements are partly shown in the following tables. The standard deviations were: 4.3 nT for A, 4.6 nT for B, and 3.4 nT for C.

P_a	1.006 640	P_b	0.004 537	P_c	0.003 385
Q_a	0.014 154	Q_b	1.006 605	Q_c	0.023 918
R_a	−0.007 599	R_b	0.026 840	R_c	1.005 029

2.2.2. *Inertial Sensing System*

The inertial measurement unit consists of three rate integral calculus gyros and three accelerometers. The attitude and heading direction of the aircraft can be obtained by means of computer processing of the angular velocity and the acceleration detected by this unit. The accuracy of roll angle, pitch angle and heading direction were as follows:
Roll angle: $\pm 0.1°$;
Pitch angle: $\pm 0.1°$;
Heading direction: $\pm 0.2°$.

2.2.3. *Fish-eye Camera*

For the distortion of the lens, the following values measured by a collimator were used:

θ^s	R_s (mm)	θ_s	R_s (mm)
0°	0.0	50°	6.783
10°	1.383	60°	8.101
20°	2.755	70°	9.389
30°	4.115	80°	10.716
40°	5.459	90°	12.056

where θ_s denotes the azimuth distance (or altitude) and R_s the distance from the center of the film.

For the installation of the fish-eye camera on the aircraft, the ideal case is when the horizontal reference plane of the inertial measurement unit and the camera are parallel to each other and the mark on the film of the azimuth of the fore-and-aft axis is also parallel to the fore-and-aft axis of the aircraft. Actually, however, they were slightly misaligned. The inclination of the fish-eye camera with reference to the horizontal reference plane of the inertial measurement unit is called the 'attachment angle' of the fish-eye camera, which can be measured with a clinometer. The deviation of the fore-and-aft azimuth mark can be measured while the aircraft is on the apron. The order of the above-mentioned measurements is as follows:

(a) The inclination of the fish-eye camera with reference to the inertial measurement unit is measured with a clinometer to obtain the attachment angle of the fish-eye camera.

(b) A theodolite is set up on the extension of the fore-and-aft axis of the aircraft on the apron, and the azimuth of the heading of the aircraft is obtained from an observation of true north using the Sun.

(c) A photograph of the Sun is taken with the fish-eye camera. The time of photographing is accurately recorded.

(d) The film of the photograph of the Sun is developed and then placed on a film-reader. With reference ot the fore-and-aft azimuth mark on the film, the altitude and azimuth of the Sun are read. From these values the correction to the attachment angle can be derived.

(e) The altitude and azimuth of the Sun at the time of photographing are obtained from astronomical calculations. From those results, together with the results of (d) above, the azimuth of the fore-and-aft azimuth mark on the film is obtained.

(f) By comparing the results of (b) and (e) above, the deviation of the fore-and-aft azimuth mark of the fish-eye camera is determined.

According to the results of the actual measurements, the inclination of the attachment angle of the camera had maximum values of 13′ for the roll angle and 8′ for the pitch angle, and the deviation of the fore-and-aft azimuth mark had a maximum value of 55′.

2.3. Observation of the Aircraft's Magnetic Field

Correction for the magnetic field of the aircraft is the most difficult problem in aeromagnetic surveys. In this survey, the aircraft used was a YS–11. The aircraft's magnetic field was measured prior to the survey each year. The measurement was carried out by calibration flights along eight passes over a check point. As the check point, the aeronautical beacon station at Niigata was selected for the proton magnetometer; while, for the three-component airborne magnetometer, the aeronautical beacon station at Shiono Misaki and the nautical lighthouse at Daiozaki were used.

The scatter of the measured values was comparatively small for the proton magnetometer. For the three-component airborne magnetometer, however, the scatter was much larger, amounting to about 100 nanotesla for each component. Like the three-component magnetometer, the fish-eye camera was mounted in the fore part of the fuselage, while the inertial measurement unit was in the after cabin and the sensor was in the tail-stinger behind the fuselage. Hence, it is considered that torsion and flexion due to vibration of the aircraft during the flight were the main cause of the scatter of these measured values.

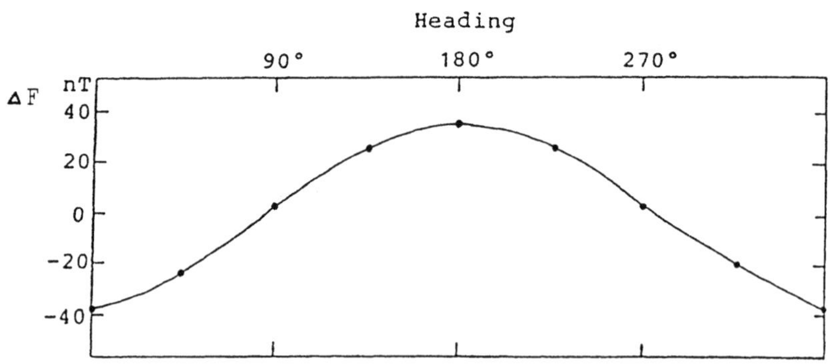

The values used in corrections for the aircraft's magnetic field were obtained from smooth curves drawn manually by connecting these measured values. In this paper, only the results of the measurements obtained using the proton magnetometer are shown in the following figure.

2.4. Data processing

The data processing was carried out entirely by a computer and a plotter. All the software used was specially developed. The data obtained from the survey were recorded on magnetic tape cassettes. The data were then processed on the large ACOS system 700 electronic computer. The substance of the data processing by the computer includes: all corrections for instrumental errors, corrections for the aircraft's magnetic field and corrections for roll and pitch of the aircraft, as well as astronomical computations to obtain the azimuth and altitude of the Sun, computation of the azimuth of the aircraft's heading, computation of the seven geomagnetic elements, corrections for altitude, etc. The computed results were recorded on magnetic tapes and printed out as lists of measured values.

Some of the equations used for the data processing are described hereunder:

(1) The corrections for instrumental errors of the fluxgate magnetometer have already been discussed in section 2.2. From Equation (1), the magnetic components, A, B, and C, along the directions of axes, a, b, and c, can be obtained from the measured values, A', B', and C' of the fluxgate magnetometer as follows:

$$\begin{aligned} A &= G_a/G \\ B &= G_b/G \\ C &= G_c/G \end{aligned} \tag{2}$$

where

$$G = \begin{pmatrix} P_a & Q_a & R_a \\ P_b & Q_b & R_b \\ P_c & Q_c & R_c \end{pmatrix}, \qquad G_a = \begin{pmatrix} A' & Q_a & R_a \\ B' & Q_b & R_b \\ C' & Q_c & R_c \end{pmatrix},$$

$$G_b = \begin{pmatrix} P_a & A' & R_a \\ P_b & B' & R_b \\ P_c & C' & R_c \end{pmatrix}, \qquad G_c = \begin{pmatrix} P_a & Q_a & A' \\ P_b & Q_b & B' \\ P_c & Q_c & C' \end{pmatrix}.$$

(2) Corrections for heading direction, roll angle and pitch angle.

If we assume the heading direction of the aircraft to be H, the roll angle R (positive as the left wing rises) and the pitch angle P (positive as the aircraft's nose rises), the coordinates obtained when corrections for H, R, and P are made to the abc coordinates fixed in the aircraft are the usual xyz coordinates. Therefore, if we assume the measured values of the magnetic components in the directions of axes a, b, and c, to be A, B, and C, respectively, the magnetic components X, Y, and Z in axes x, y, and z, respectively, are expressed as follows:

$$\begin{pmatrix} X \\ Y \\ Z \end{pmatrix} = \begin{pmatrix} \cos H \cos P & -\sin H \cos R + \cos H \sin P \sin R & \sin H \sin R + \cos H \sin P \cos R \\ \sin H \cos P & \cos H \cos R + \sin H \sin P \sin R & \cos H \cos R + \sin H \sin P \sin R \\ -\sin P & \cos P \sin R & \cos P \cos R \end{pmatrix} \begin{pmatrix} A \\ B \\ C \end{pmatrix}. \tag{3}$$

Geomagnetic elements such as D, I, H, and F can be obtained easily from X, Y, and Z.

(3) Corrections for altitude.

The flight altitude during the survey was 9500 feet. In order to reduce the observed values to those at the sea level, corrections for altitude were made using the gradients of the main geomagnetic field. The MGST (6/80) model of the main magnetic field was used to calculate these gradients.

3. Geographical Distribution of Magnetic Elements

The geographical distribution of the magnetic elements was determined using data obtained from standard magnetic observatories, land magnetic surveys and aeromagnetic surveys. The equations for the average distribution of the Earth's magnetic field components in the vicinity of Japan were approximated by third-degree polynomials of latitude and longitude.

If we assume

$$\Delta\phi = \phi - 34^\circ\,\mathrm{N}, \qquad \Delta\lambda = \lambda - 136^\circ\,\mathrm{E}$$

where ϕ is latitude, λ longitude, and the origin of the coordinates is at 34° N, 136° E, the distribution of each element of the geomagnetic field can be expressed by the following equation (Units for $\Delta\phi$ and $\Delta\lambda$ are minutes):

$$M = a + b\Delta\phi + c\Delta\lambda + d\Delta\phi^2 + e\Delta\phi\Delta\lambda + f\Delta\lambda^2 + g\Delta\phi^3 + h\Delta\phi^2\Delta\lambda + \\ + i\Delta\phi\Delta\lambda^2 + j\Delta\lambda^3.$$

Thus, for the components X, Y, and Z,

$$\begin{aligned} X &= a_1 + b_1\Delta\phi + c_1\Delta\lambda + \ldots + i_1\Delta\phi\Delta\lambda^2 + j_1\Delta\lambda^3 \\ Y &= a_2 + b_2\Delta\phi + c_2\Delta\lambda + \ldots + i_2\Delta\phi\Delta\lambda^2 + j_2\Delta\lambda^3 \\ Z &= a_3 + b_3\Delta\phi + c_3\Delta\lambda + \ldots + i_3\Delta\phi\Delta\lambda^2 + j_3\Delta\lambda^3. \end{aligned} \tag{4}$$

Each coefficient of Equations (4) can be obtained from the survey results by means of simple least squares. The results are as shown in the following table. The distribution equations for declination (D), dip (I), horizontal intensity (H) and total intensity (F) were obtained from the distribution equations for X, Y, and Z. The geographical distribution of the seven elements of the Earth's magnetic field for 1980.0 in the seas around Japan are shown in Figures 2–8.

	a	b	c	d $\times 10^{-3}$	e $\times 10^{-3}$	f $\times 10^{-3}$	g $\times 10^{-6}$	h $\times 10^{-6}$	i $\times 10^{-6}$	j $\times 10^{-6}$	Probable error
D (West)	6°09.8′	0.3294′	−0.0795′	0.0114′	0.0367′	−0.1381′	0.0320′	−0.0326′	0.0187′	−0.0095′	±9.1′
I	46°51.1	1.3448	−0.1869	−0.3147	−0.0234	0.0039	−0.0174	0.1550	−0.0633	0.0573	±7.3
	nT	nT	nT	nT	nT	nT	nT	nT	nT	nT	nT
H	31454	−6.7870	−1.4251	−1.4037	2.1362	−0.4348	0.6887	−0.5532	0.0536	−0.977	±78
F	45993	9.2192	−4.7219	0.7263	−0.4394	−0.2823	−1.0741	1.1490	−0.5136	0.5786	±60
X	31273	−7.0744	−1.3343	−1.4849	2.1548	−0.3158	0.7049	−0.6048	0.1806	−1.1451	±79
Y	−3377	−2.2720	0.8804	0.7109	−0.5891	1.2745	−0.1905	0.1395	−0.2748	0.0000	±84
Z	33556	19.0007	−5.1400	−2.4848	−1.2782	−0.0638	−0.8067	2.6096	−0.9545	0.9122	±86

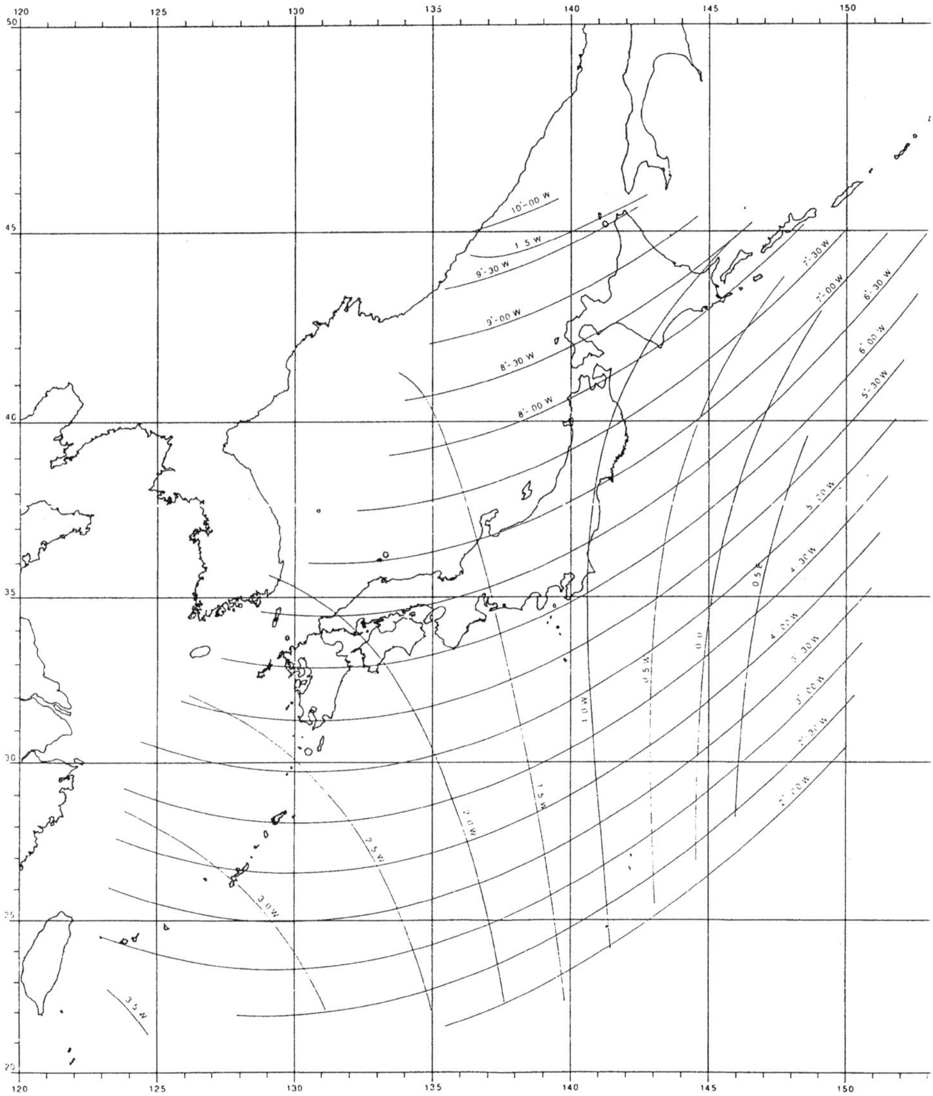

Fig. 2. Lines of Equal Magnetic Declination (δ) and of Equal Annual Change for 1980.0.

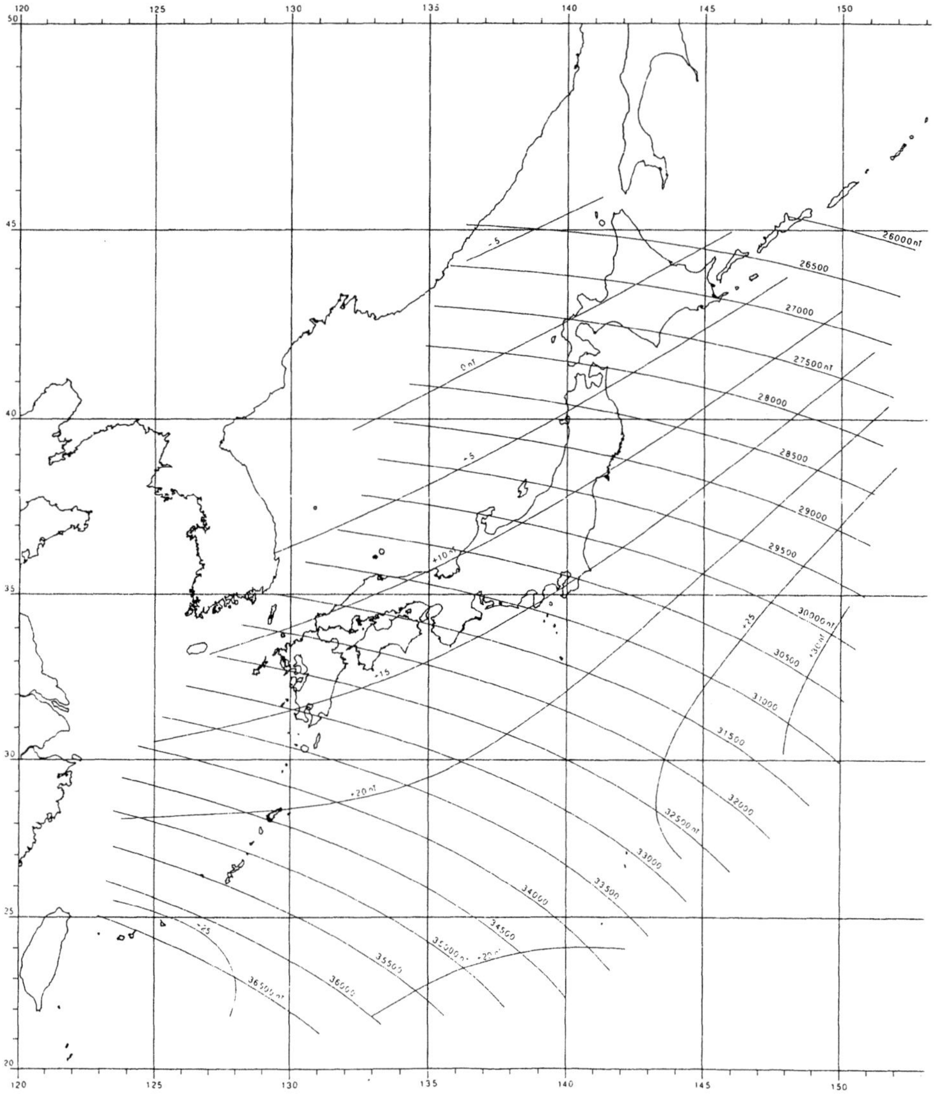

Fig. 3. Lines of Equal Horizontal Intensity (H) and of Equal Annual Change for 1980.0.

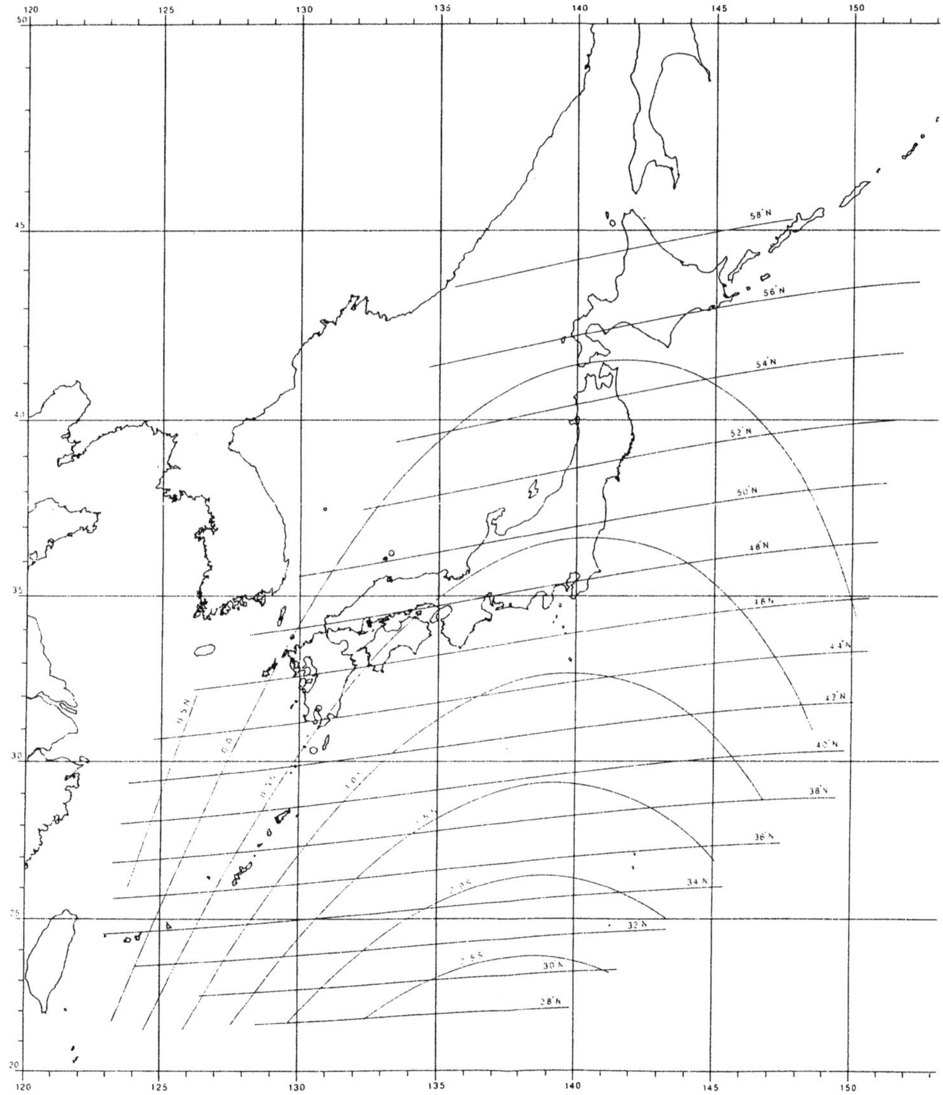

Fig. 4. Lines of Equal Magnetic Dip (θ) and of Equal Annual Change for 1980.0.

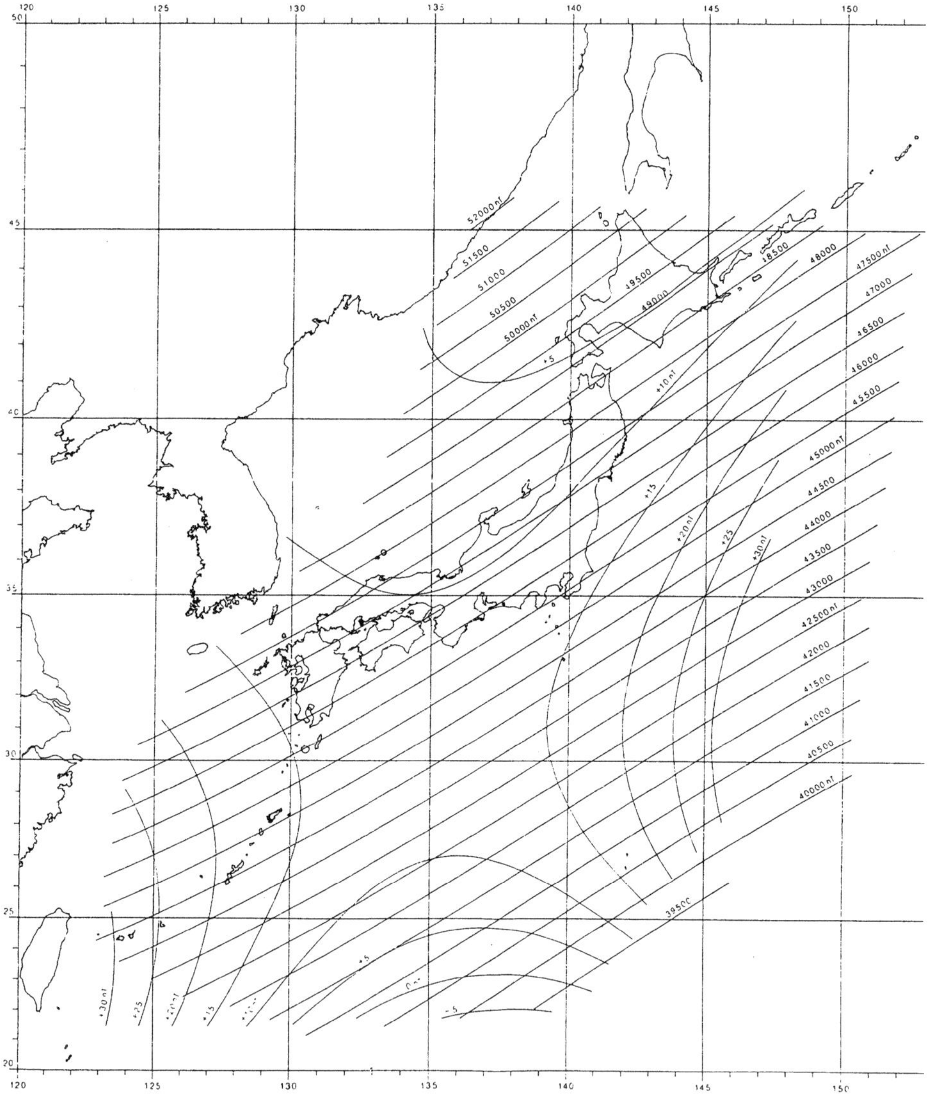

Fig. 5. Lines of Equal Total Intensity (F) and of Equal Annual Change for 1980.0.

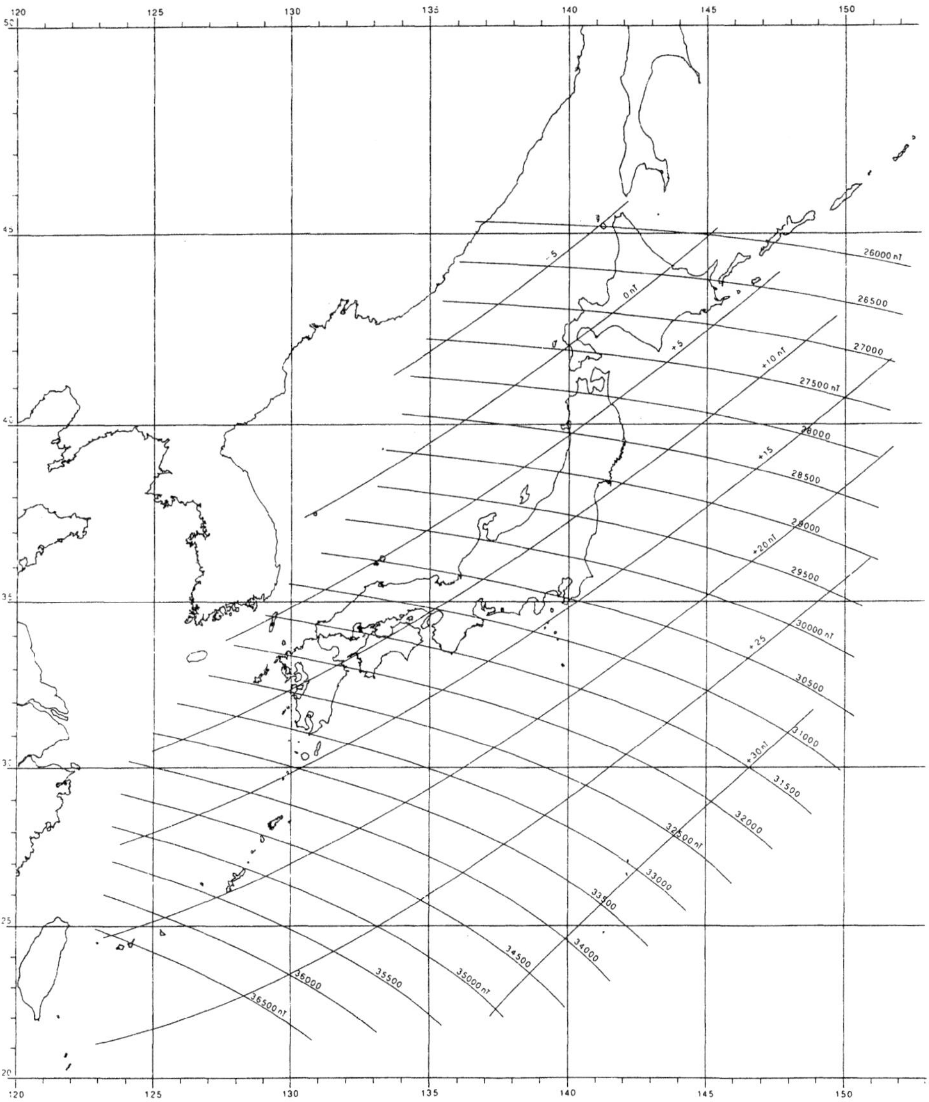

Fig. 6. Lines of Equal North Component (X) and of Equal Annual Change for 1980.0.

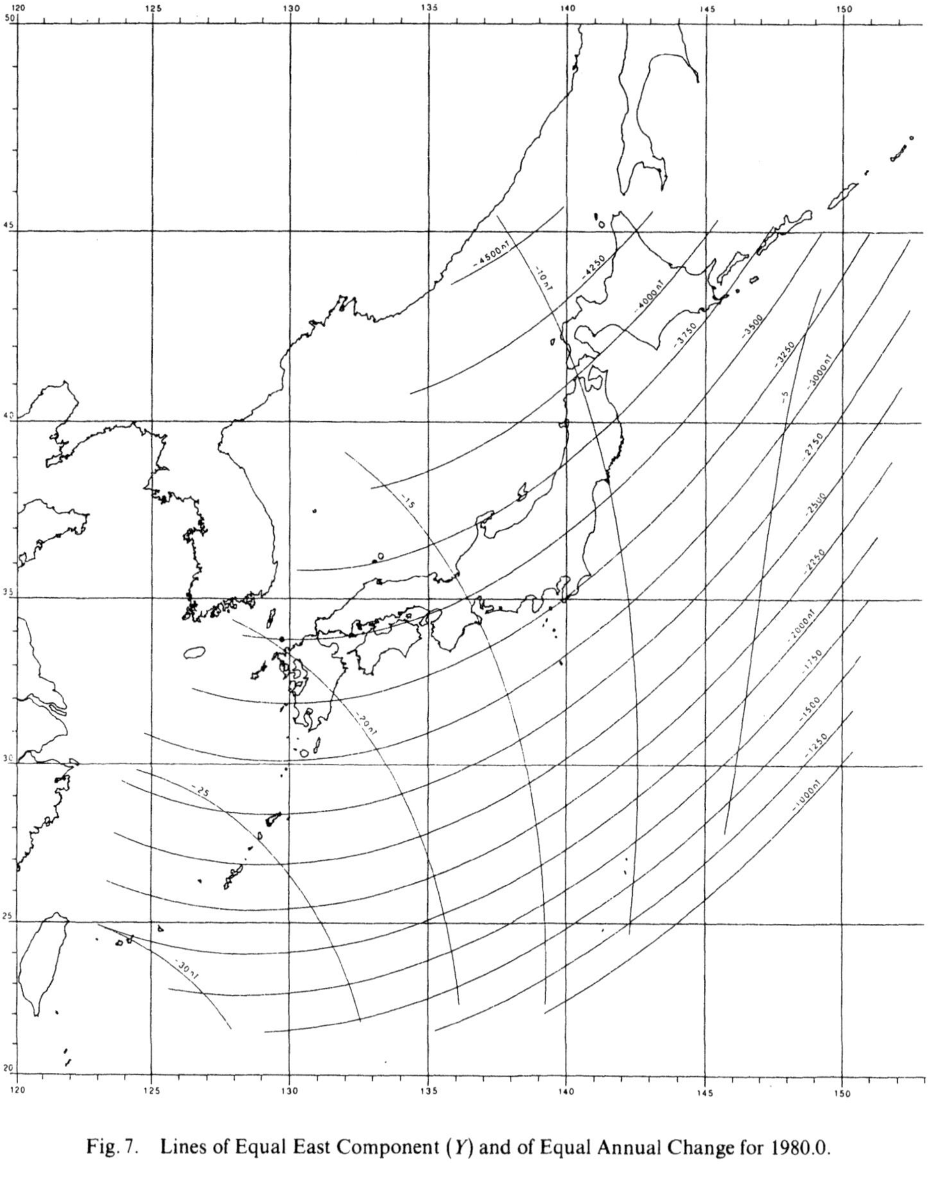

Fig. 7. Lines of Equal East Component (Y) and of Equal Annual Change for 1980.0.

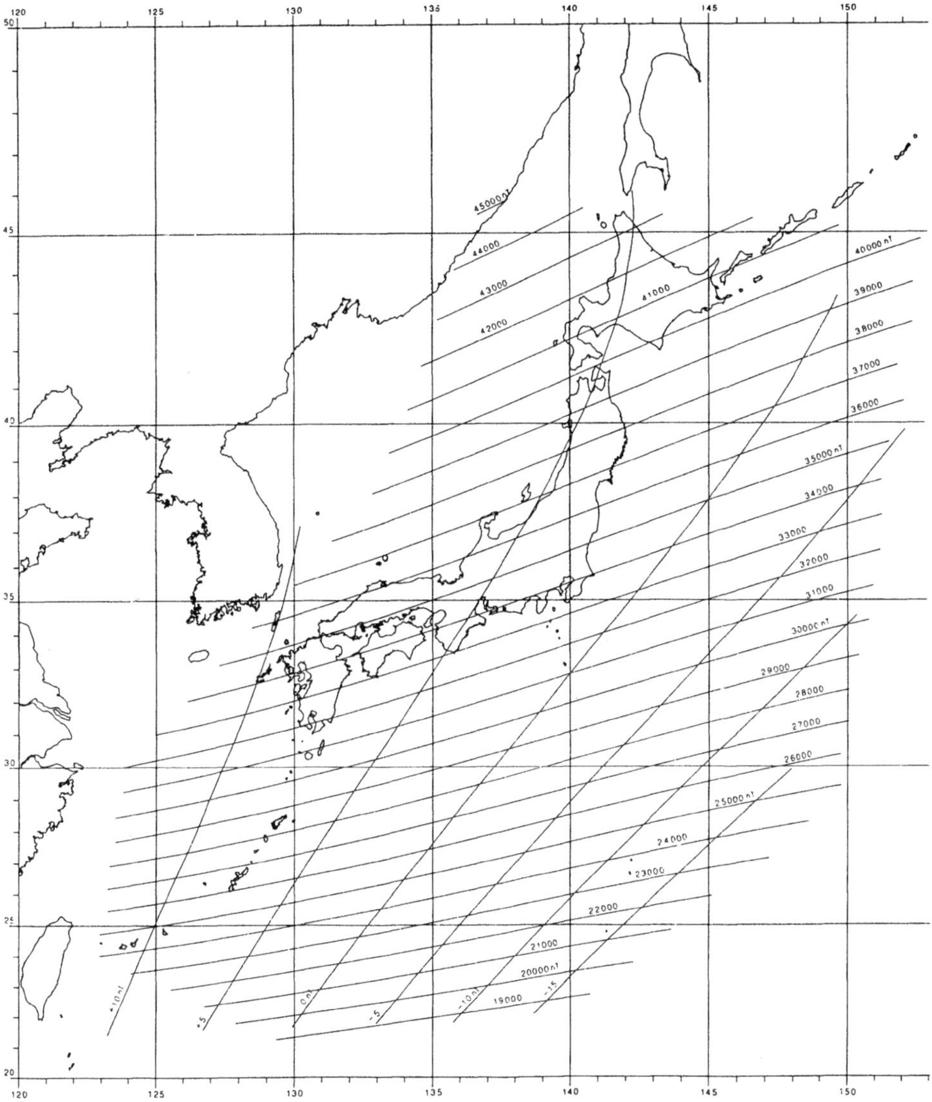

Fig. 8. Lines of Equal Vertical Intensity (Z) and of Equal Annual Change for 1980.0.

4. Secular Variation in and Around Japan

The distribution of the predicted values of annual change were obtained by using the data observed up to 1980 at 14 observatories in Japan and neighbouring countries. Estimation of the secular variation can be made either by the method of extrapolation using an approximation equation for the variation, or by the method of freehand drawing to derive future values of the variation by plotting the data on a graph.

In this paper, the latter method was used, because the secular variation around Japan in recent years has exhibited inflection points. The distribution of annual changes, m, is expressed as follows:

$$m = a + b\Delta\phi + c\Delta\lambda + d\Delta\phi^2 + e\Delta\phi\Delta\lambda + f\Delta\lambda^2 + g\Delta\phi^3 + h\Delta\phi^2\Delta\lambda + \\ + i\Delta\phi\Delta\lambda^2 + j\Delta\lambda^3$$

where $a, b, \ldots j$ are the constants to be determined by the method of least squares, and $\Delta\phi$ and $\Delta\lambda$ have been defined above. The observatories whose data were used in the computation are as follows:

Memambetsu, Kakioka, Kanoya, Mizusawa, Kanozan, and Chichijima in Japan; Yuzhno-Sakhalinsk, Vladivostok, Magadan, Yakutsk, Irkutsk, and Petropavlovsk in the U.S.S.R.; Guam in the Marianas Islands; Mutinlupa in the Philippines.

The coefficients in the distribution formulae for the annual changes are as shown in the following table:

	a	b $\times 10^{-3}$	c $\times 10^{-3}$	d $\times 10^{-6}$	e $\times 10^{-6}$	f $\times 10^{-6}$	g $\times 10^{-9}$	h $\times 10^{-9}$	i $\times 10^{-9}$	j $\times 10^{-9}$	Probable error
$\frac{dD}{dt}$ (West)	1.7′	−0.8′	−1.8′	0.5′	2.5′	−1.9′	0.6′	1.2′	0.8′	−0.9′	±0.1′
$\frac{dI}{dt}$	−0.7	2.2	−1.1	−0.9	0.2	2.2	0.1	0.6	−2.0	0.0	±0.1
$\frac{dH}{dt}$	14nT	−30nT	12nT	−17nT	12nT	5nT	24nT	−29nT	−4nT	4nT	±1nT
$\frac{dF}{dt}$	11	−12	6	−21	−2	32	37	−43	−30	17	±1
$\frac{dX}{dt}$	12	−33	17	−7	5	3	9	−10	3	−1	±1
$\frac{dY}{dt}$	−16	8	13	1	−14	7	−3	3	−6	−1	±1
$\frac{dZ}{dt}$	5	11	−14	−12	18	−5	1	−7	1	0	±1

5. Conclusions

The magnetic charts in this paper have been compared with those for the IGRF for 1980 [4]. It was found that these charts are very similar to the IGRF charts, as shown in Figure 9.

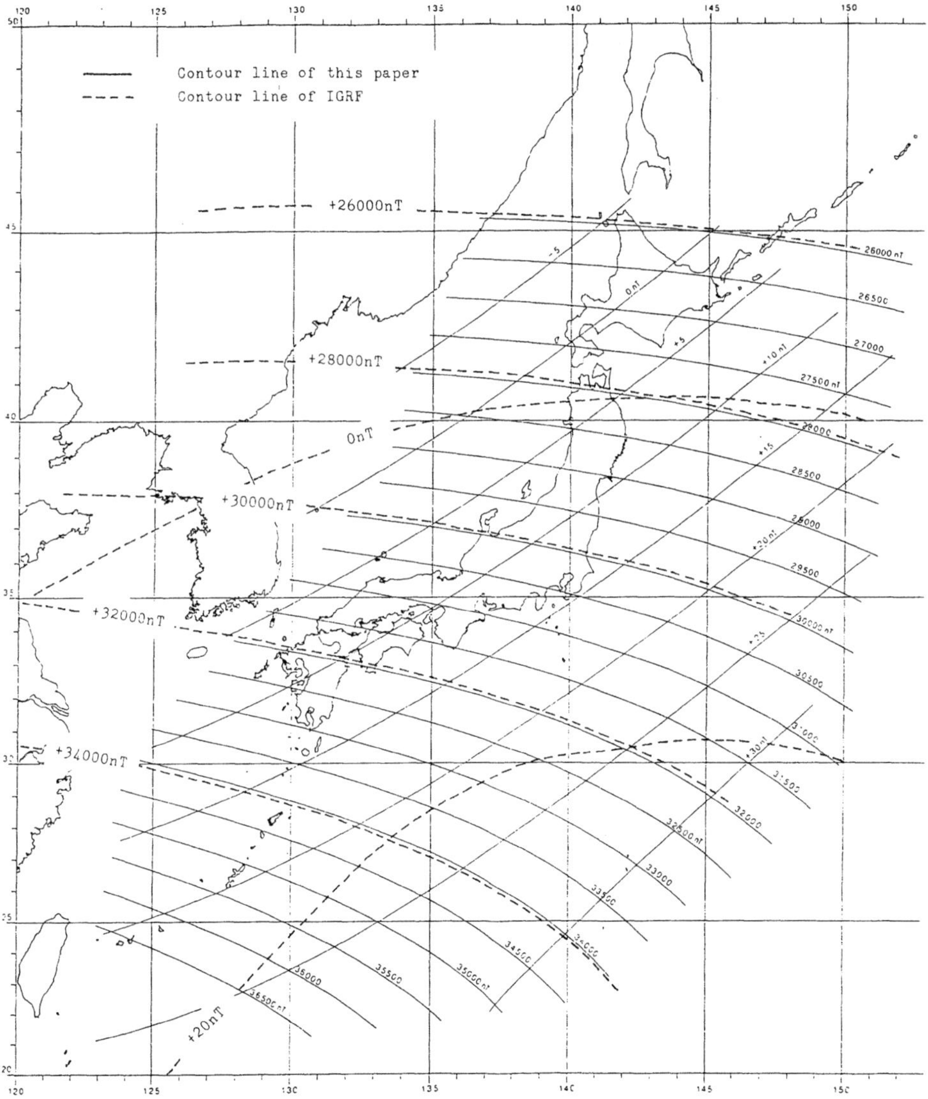

Fig. 9a. Comparison with the IGRF charts for 1980. *X* Component.

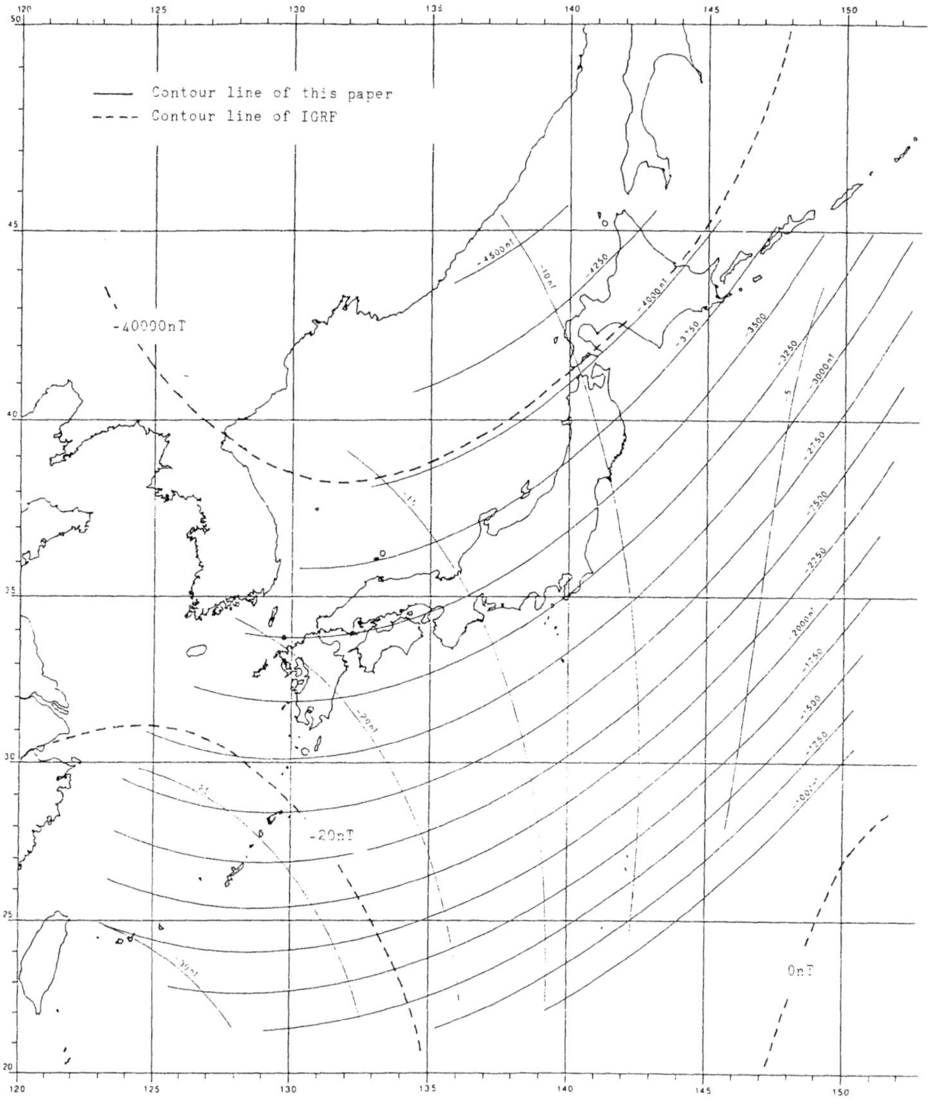

Fig. 9b. Comparison with the IGRF charts for 1980. *Y* Component.

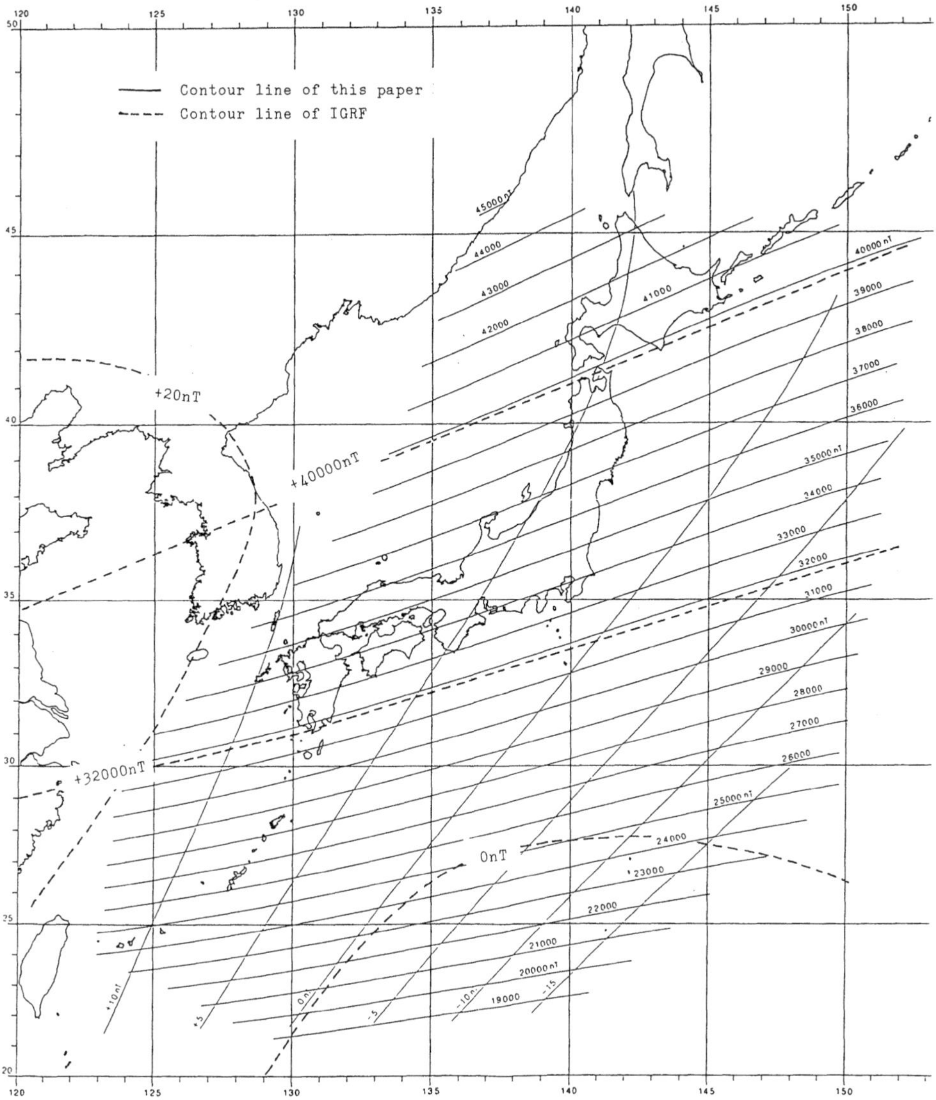

Fig. 9c. Comparison with the IGRF charts for 1980. *Z* Component.

Acknowledgments

The authors are grateful to the members of the Haneda Air Base, the 3rd Regional Maritime Safety Headquarters, for operation of aircraft. The authors wish to express their sincere thanks to staff members of the Kakioka Magnetic Observatory of the Japan Meteorological Agency for their kind cooperation.

References

[1] World Data Center C2 for Geomagnetism: 1966, Report on Aeromagnetic Survey in Japan, pp. 1–242.
[2] Hydrographic Department, M.S.A., Japan: 1973, 'Magnetic Survey of Japan 1969–1970', Data Report of Hydrographic Observations, *Series of Magnetic Survey*, No. 2, March.
[3] Hydrographic Department, M.S.A., Japan: 1978, 'Magnetic Survey of Japan, 1974–1975', Data Report of Hydrographic Observations, *Series of Magnetic Survey*, No. 3, June.
[4] Fabiano, E. B., Peddie, N. W., Barraclough, D. R., and Zunde, A. K.: 1982, 'International Geomagnetic Reference Field 1980, Charts and Grid Values', *IAGA Bulletin* No. 47, Geological Survey Circular 873, U.S.A., pp. 1–142.

AN AUTOMATIC AUDIO-MAGNETOTELLURIC EQUIPMENT, CONTROLLED BY MICROPROCESSOR, FOR THE TELESURVEILLANCE OF THE VOLCANO MOMOTOMBO (NICARAGUA)

G. CLERC, J-P. DÉCRIAUD, G. DOYEN, M. HALBWACHS*, M. HENROTTE, J. RÉMY, and X-C. ZHANG**

Centre de Recherches Géophysiques de GARCHY 58-150 Pouilly-s-Loire, France

Abstract. A campaign of audio-magnetotelluric soundings in the crater of Momotombo has shown that the structure of this volcano is suitable to a surveillance by this method; it has an only central activity, and the resistivity versus the depth decreases suddenly at least 100 times at about 270 m.

So, this technical paper describes the equipment designed and built in order to measure every 2 hr the apparent resistivity in a given direction, at seven frequencies regularly spaced from 5 to 312 Hz.

The data are preprocessed to fill only 32 bytes: they are

- on the one hand, printed locally for the close surveillance;
- on the other hand, transmitted by an ARGOS platform and received in Garchy via satellite for searching studies.

The system is now undergoing technological and environmental trials on an italian volcano. Geophysical results will form the matter of a later on paper.

1. Introduction

In volcanic or geothermal areas, the hot parts of the substratum are generally very conductive so that the magnetotelluric (M-T) method is well suited for prospecting them. It is actually in current use among scientific searchers (cf.: bibliography) and exploration geophysicists.

It will be seen hereafter how the results of measurements performed on a volcano, the structure of which is appropriate, have led us to use this method for a permanent surveillance.

Then, the fully automatic system designed for this purpose will be described as also the program according to which data acquisition and transmission are performed.

Geophysical results have not yet been get: they will form the matter of another paper later on.

2. Results of Previous Non Automatic Measurements

In December 1981, seven audiomagnetotelluric (AMT) soundings were carried out at the bottom of the crater of the volcano Momotombo (Nicaragua). The results yielded

* Université de Savoie, 73 000 Chambéry, France.

** Institute of Geology, Changchun, China.

Geophysical Surveys **6** (1984) 291-304. 0046-5763/84/0063-0291$02.10

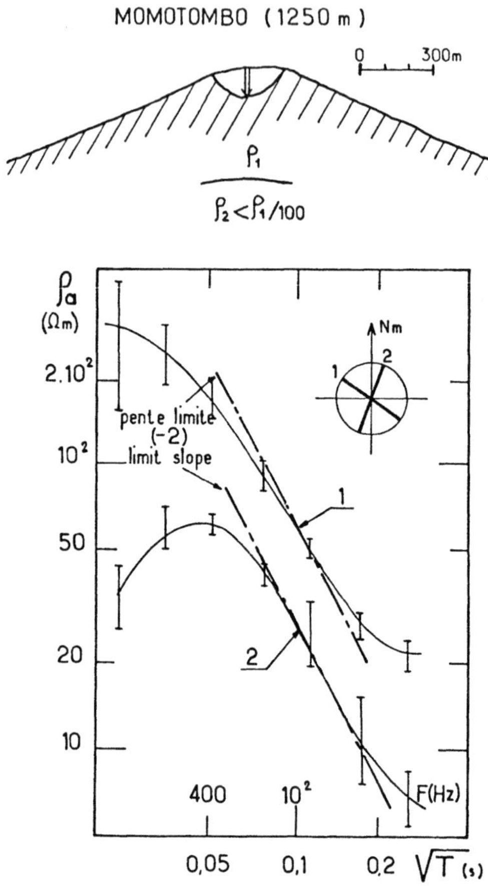

Fig. 1 Results of measurements performed in December 1981 in the crater of the volcano Momotombo (M. Halbwachs).

by two soundings performed in two particular directions are displayed on the Figure 1, according to the Cagniard's scalar formula

$$\rho_a = 0.2T\left(\frac{Ex}{Hy}\right)^2 \qquad (\Omega\text{m, s, mV/km, n}T)$$

where

ρ_a = the apparent resistivity at the period T,
Ex = the natural electric field in the direction of the telluric line,
Hy = the associated magnetic field in the orthogonal direction.

In the first case (curve 1) the telluric line was placed along the NW/SE direction, in fact 310°, corresponding approximately to the Marrabios range (300 to 305°). In the second case (curve 2), the telluric line was lying at 10°, parallel to the 1905 outflow, that is in the direction of the outfall of the crater.

These both soundings were well reproduceable so that the following remarks can be expressed: if we assume a tabular medium, a very resistive layer lying on a very conductive substratum yields a resistivity curve the slope of which is -2 on the usual bilogarithmic $\rho_a(\sqrt{T})$ diagram (Dupis, 1971). Now the both curves of the Figure 1 comprise a straight part, the slope of which is about -2: that means the contrast of resistivity is at least 100. The curve (1), get rather far from the overflow of the crater is less subject to lateral effects than the other one: it yields 200 Ωm as a mean value of the resistivity of the upper layer. Then, the resistivity of the substratum would be about 2 Ωm, or even less.

The resistivity of some very weathered andesitic rocks can sometimes be as low as 10 Ωm, but we do not know that values as low as a few Ωm only have been found elsewhere than in hot areas. The depth of the top of this conductive layer is yielded by the formula $p = 356\sqrt{\rho_a T}$ (m, Ωm, s) (Dupis, 1971). ($\rho_a T$ is constant in the part of the curve the slope of which is -2).

The difference between the curves (1) and (2) shows the 1–D interpretation is inadequate. In the present case, however, the structural models point out that a rather accurate depth is get from the sounding with the telluric line parallel to the direction of anisotropy (curve 1), that is

$$p \simeq 270\,\mathrm{m}.$$

Without interpreting conclusively the nature of this very conductive body (magmatic intrusion, mineralized hot water, ...), it can be thought it is directly connected to the activity of the volcano. Then, the hot front corresponding to the top of the conductive body at the depth p might go up or down according as this activity is increasing or decreasing.

The idea of measuring continuously the value of the parameter p seems the more interesting as the temperature of the fumarolean areas of the crater is increasing since several years (700 to 800 °C in 1981). Moreover, the volcanic structure of Momotombo is perfectly conical with an only central activity: a single automatic unit put in the crater and measuring the apparent resistivity ρ_a at some frequencies corresponding to the steepest part of the curve $\rho_a(\sqrt{T})$ might allow that.

The expression of p points out that a significative variation of 10% for ρ_a at a given frequency would mean a variation of 5% for p. The resolution capacity of the method is naturally limited by the scattering of the results of the measurements, but their redundancy and an appropriate processing will reduce this scattering and eliminate some probable cyclic effects (such as the diurnal rotation around the earth of the principal sources of the natural electromagnetic field in the audio-range).

3. Description of the Built Equipment

The main part of this equipment is an automatic data acquisition unit, the synoptic scheme of which is displayed Figure 2. It comprises principally:

- the sensors: magnetometer and electrodes,

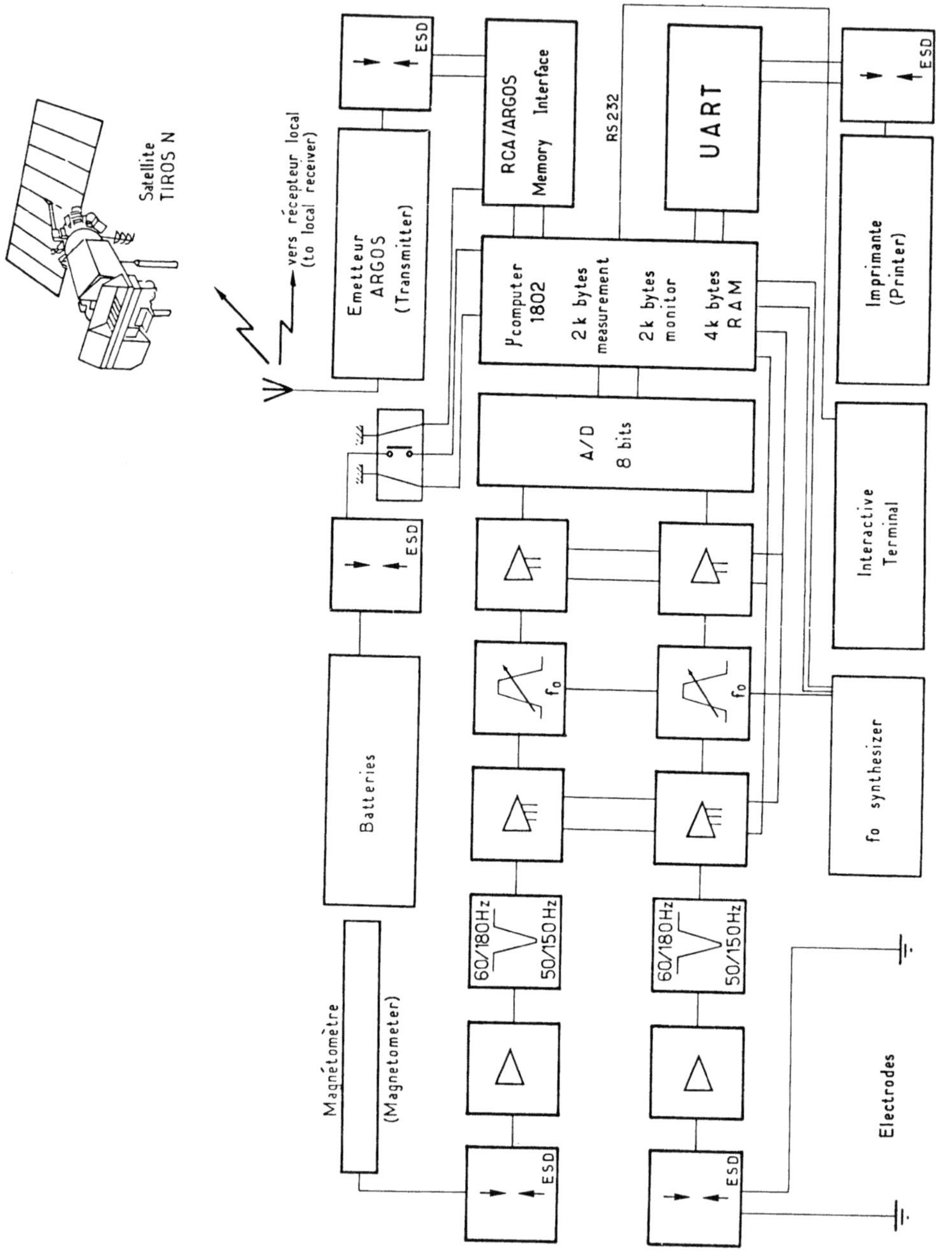

Fig. 2. Synoptic scheme of the acquisition unit.

– the properly so called acquisition system, composed of:
  . analog circuits with programmable-gain preamplifiers and filters;
  . an analog to digital (A/D) converter;
  . digital circuits with a RCA 1802 microprocessor managing the acquisition and conversion processes, and delivering the data to a built-in memory interface and to the following peripherics:
– a printer,
– an ARGOS transmitter,
– an interactive terminal (only for the tests).

The whole is supplied by two 12V/45Ah batteries providing one month of operating.

Devices protect every sub-assembly against electrostatic discharges (ESD) which could be induced in the interconnection cables during atmospheric storms. The photography of the Figure 3 shows the different components of the acquisition unit (artificially gathered together for the picture).

Every case is made of PVC and sealed because of the corrosive emanations of the volcano.

Without entering in all the details, the following remarks can be stated about the equipment:

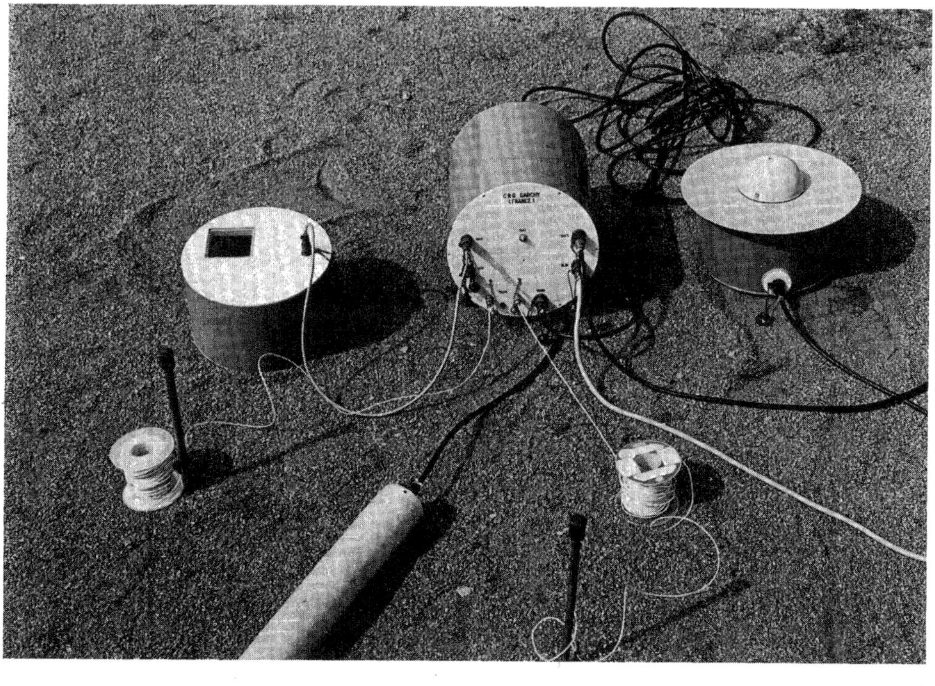

Fig. 3. General view of the acquisition and transmission equipment to be set up at the top of the volcano (gathered for the photography). In the upper part, from left to right: the printer, the acquisition system and the ARGOS transmitter. In the lower part: the magnetometer between the 2 electrodes and wires of the telluric line.

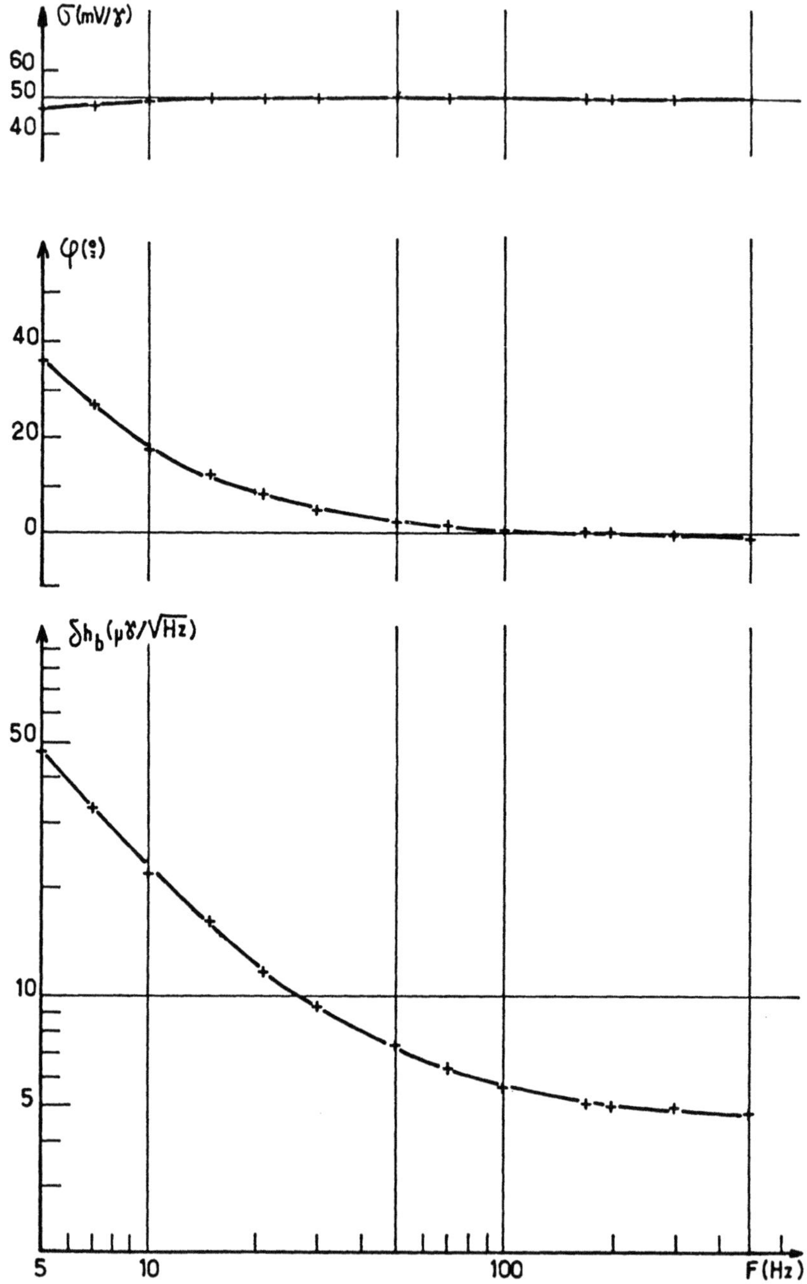

Fig. 4. Response of the magnetometer versus the frequency: σ: amplitude. φ: phase. δh_b: noise spectral density.

(1) The magnetometer is an inductive one, with a flux-feedback, in order to have a constant amplitude response versus the frequency, and stable in course of time: it was built in the C.R.G.

Its specifications are given on the Figure 4.

(2) The electrodes are made of an anticorrosion alloy (ARCAP), but their replacement is provided for, if necessary. The telluric line is 50 m long.

(3) In the analog part, the signals are amplified by a fixed-gain stage in each channel and then interferences at the mains and third harmonic frequencies are removed by notch-filters for avoiding to overload the following stages. These comprise:

- First, a wide-band amplifier the gain of which is automatically chosen by the μP versus the amplitude of the wide-band signal to be measured.
- Then, a switched-capacitors bandpass filter: the tuning frequency is one hundredth of the frequency f_0 synthetized by a circuit controlled by the μP which elaborates so seven successive frequencies.
- At last, another amplifier the gain of which is chosen by the μP to, versus the amplitude of the narrow-band signal delivered by the filter, in order to be large enough for the 8 bits A/D converter.

(4) The digital part of the unit comprises three cards from RCA: (CPU, A/D converter and UART) and a specific RCA/ARGOS Memory interface.

As for the peripherics:

– The local printer is a 5020 OEM model made by KONTRON with a NIP 18 mechanism from OLIVETTI: it is connected to the RS 232 output of the UART.

– The transmitter is an ARGOS VHF platform sending during about 1 s every 200 s, a message containing 32 significative bytes.

– An any interactive terminal can be connected provisionally to a second RS 232 output without removing the printer: it is very useful for setting the system right and for further checking.

4. Data Acquisition and Preprocessing Program

The chart-flow of the Figure 5 shows this program.

Between two measurements, the supply of the acquisition unit is switched off to increase its duration. Only the pilot-oscillator of the ARGOS transmitter is running as also the dividers controlling

- on the one hand the measuring rate (every 2 hr);
- on the other hand the transmitting rate (every 200 sec)

and the RCA/ARGOS memory interface.

So, the initialization of a measurement begins by waking up the system: it starts with maximum values of the programable gains and of the tuning frequency ($f_0/100 = 312$ Hz).

4.1. Calibration at 312 Hz

Calibrating signals at the same frequency are applied:

- on the one hand to the calibrating coil of the magnetometer;
- on the other hand to the input of the telluric channel.

These signals have such an amplitude that all the programable gains have to change to their minimum values for avoiding any overload: this test checks not only the

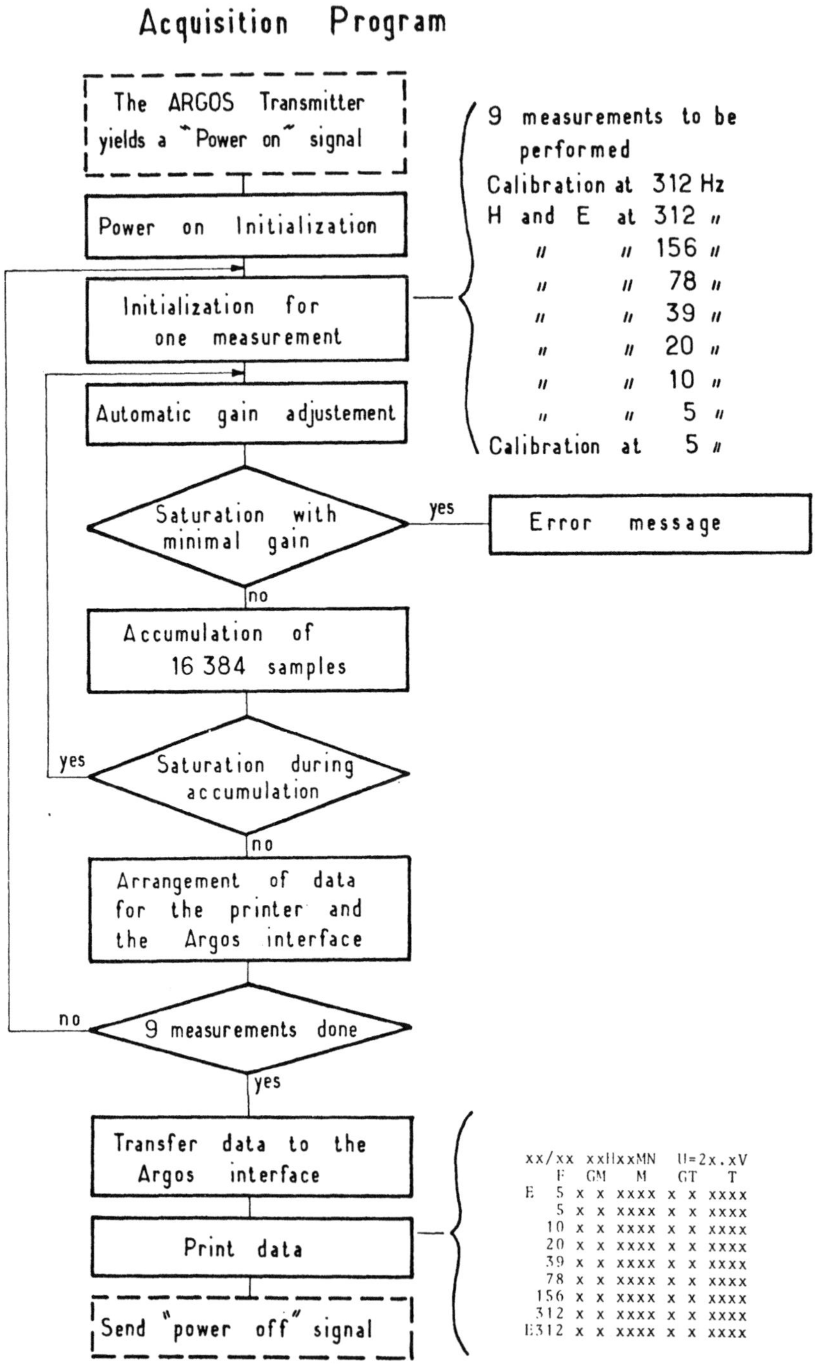

Fig. 5. Data acquisition and preprocessing program.

stability of the both channels, but the good running of the automatic control of the gains too.

The principle of the preprocessing is to get a mean value of the energy of the signal on each channel during a given time.

For doing that, 64 successive accumulations of 256 samples, that is 16 384 samples on each channel are added and averaged, and the result is stored into a memory.

4.2. Measure at 312 Hz

Then the calibrating signals are switched off and the gains set at their maximum values again, in order to measure the natural signals at the same frequency (312 Hz). The gains settle themselves to the maximum possible values without overloading the amplifiers: if the fluctuations of the natural signal lead one of them to be overloaded, the gain is adjusted again and accumulation restarts from the beginning.

However, precautions are taken for avoiding that occasional short interferences oblige to start the accumulation again, because their weight in the result is very weak in any case. When the accumulation is over, the results of both channels are stored as also the corresponding values of the gains.

4.3. Measures at the other frequencies and calibration at 5 Hz

Then, the same process takes place at the other designed frequencies and a calibration at the lowest frequency (5 Hz) is performed too.

4.4. Monitoring of the battery

At last, the voltage of the battery in monitored through the A/D converter.

4.5. Data output

All the data are then moved to the memory of the RCA/ARGOS interface as also to the local printer which is switched on only during the minimum time for printing the 11 lines of the message. The appearance of the printed message can be seen in the right lower part of the Figure 5.

The distribution of the data into the 32 bytes of the ARGOS message is as follows:

– a message number to be replaced later by the day number and the hour (to 1 min)		4 bytes
– the voltage of the battery after the measurement (to 0.1 V)		1 byte
– the mean values of the magnetic (M) and telluric (T) signals at 7 frequencies and 2 calibrations:	$2 \times (7 + 2) =$	18 bytes
– the corresponding gains (GM and GT):	$2 \times 1/2 \times (7 + 2) =$	9 bytes

(each gain fills 1/2 byte).

The chart flow let see the possibility of an error message which occurs when an amplifier is overloaded, even with the minimum gain: in this case, the three bytes containing the results (M, T, and GM/GT) are forced to zero, except some bits indicating what amplifier is overloaded.

4.6. Timing of one complete acquisition

Table I sums up the sampling conditions and the duration of the accumulations.

TABLE I

Sampling rate and duration of the accumulations

Frequency (Hz)	Sampling step (μs)	Number of samples per period (1 channel)	Number of samples per second (2 channels)	Duration of 1 accumulation (sec) (2 channels)
312.5	256	6.25	3906	8.39
156.2	256	12.5	3906	8.39
78.13	256	25	3906	8.39
39.06	256	50	3906	8.39
19.53	512	50	1953	16.78
9.77	1024	50	977	33.55
4.88	2048	50	488	67.15

Taking into account the time for calibrating and for setting the gains right, a full measuring cycle will take nearly 4 min, that is more than the time between two successive transmissions from the ARGOS platform. The transmission of the data of the previous measurement have priority over the measurement in progress: this one is temporarily interrupted (less than 1 sec) to avoid interferences from the transmission.

5. Data Reception

Two ways of receiving the data are designed:

(1) The way shown Figure 6: the CNES-NOAA receiving network collects in Toulouse the data from the satellites. From Garchy, a HP85 microcomputer interrogates automatically the Toulouse data centre four times a day and sends the data for storage to the 'Centre Interrégional de calcul' (C.I.R.C.E.) in Orsay, via the 'Transpac' transmission network.

This receiving system is fully operating, and data coming from Etna and South Italy are already regularly processed.

(2) The way shown Figure 7: this way is not yet ready. It aims at setting up the printer in an easily accessible building at the foot of the volcano instead of in the crater: the main difficulty consists in setting up on the edge of the crater a receiving-transmitting relay station which could resist to the corrosion.

6. Conclusion

From an instrumental point of view, the equipment works according to the expectation: it is now undergoing technological and environmental trials near Pouzzoles, in South Italy, since December 1983.

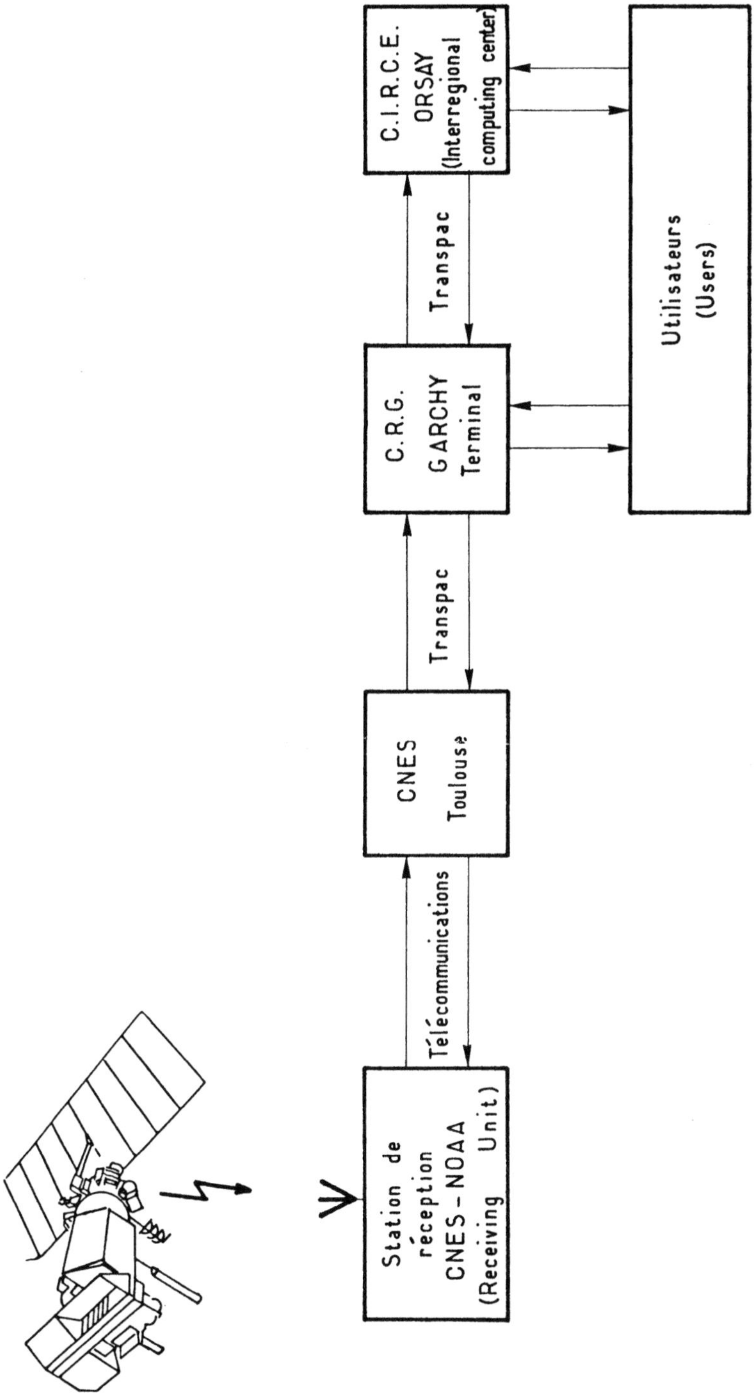

Fig. 6. Data reception at the Garchy C.R.G.

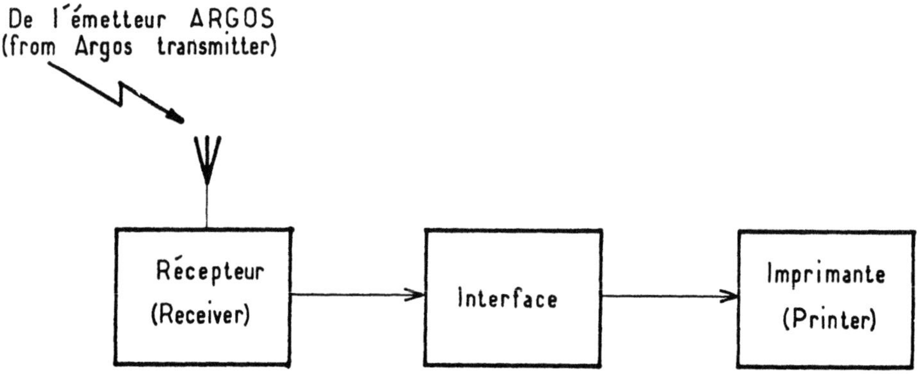

Fig. 7. Data reception at the foot of the volcano (in progress).

It can be told that the ARGOS platform, in spite of its limited capacity (32 bytes), can transmit rather sophisticated informations on condition that an appropriate pre-processing be done: the disadvantage consists in a greater consumption of current, reducing the autonomy to one month. Later on, when the printer for the local surveillance will be at the foot of the volcano, the simultaneous recharge of the batteries will be absolutely necessary. Technological trials of solar cells and air-generator are undergoing in the corrosive environment of the top of Momotombo.

From a geophysical point of view, the time to conclude has not yet come: a further paper will be necessary.

Acknowledgments

The authors express their thanks to J. Tabbagh who developed the software for interrogating the C.N.E.S. Data Centre in Toulouse and for transmitting the data to the C.I.R.C.E. in Orsay (Figure 6). This work took place in the organization of a volcanologic data centre in Garchy under the care of the P.I.R.P.S.E.V. (Programme Interdisciplinaire de Recherches sur la Prévision et la Surveillance des Eruptions Volcaniques).

References

Ander, M. E., Goss, R., Strangway, D., Hillebrand, C., Laughlin, A. W., and Hudson, C.: 1980, 'Magnetotelluric/Audiomagnetotelluric Study of the Zuni Hot Dry Rock Geothermal Prospect, New Mexico', *Geothermal Resources Council, Trans.* **4**.

Ander, M. E.: 1981, 'Magnetotelluric Exploration in Arizona and New Mexico for Hot Dry Rock Geothermal Energy Using SQUID Magnetometers', in H. Weinstock and W. C. Overton, Jr.(eds.), *SQUID Applications to Geophysics*, Soc. Explor. Geophys., pp. 61–65, Tulsa, Oklahoma.

Ander, M. E.: 1982, 'Magnetotellurics Applied to Hot Dry Rock Geothermal Exploration in Arizona and New Mexico', *Hot Dry Rock Geothermal Exploration Workshop*, Los Alamos National Laboratory.

Aubert, M., Dupis, A., Lenat, J. F., Roux, J., et Senaud, G.: 1979, 'Exploration géophysique de la fosse volcano-tectonique du Cantal', *Compt. Rend. Acad. Sci. Paris*, t. 288. Série D, 215–218.

Aubert, M., Dupis, A., Lenat, J. F., Roux, J., et Senaud, G.: 1982, 'Structure of the Cantal Strato-Volcano, French Central Massif, from Electric, Magnetotelluric Soundings and Aeromagnetic Data', *Journ. of Volcanology and Geothermal Research* **12**, 77–99.

Ballestracci, R. et Benderitter, Y.: 1980, 'Sondages magnéto-telluriques dans le rift d'Asal (République de Djibouti)', *Bull. Soc. géol. France*, (7), t. XXII, n°6, pp. 873–879.

Beblo, M. and Bjornsson, A.: 1978, 'Magnetotelluric Investigation of the Lower Crust and Upper Mantle Beneath Iceland', *J. Geophys.* **45**, 1–16.

Beblo, M., Bjornsson, A., Arnason, K., Stein, B., and Wolfgram, P.: 1983, 'Electrical Conductivity Beneath Iceland – Constraints Imposed by Magnetotelluric Results on Temperature, Partial Melt, Crust and Mantle Structure', *J. Geophys.* **53**, 16–23.

Beblo, M. and Bjornsson, A.: 1980, 'A Model of Electrical Resistivity Beneath NE-Iceland: Correlation with Temperature', *J. Geophys.* **47**, 184–190.

Benderitter, Y., and Gerard, A.: 1984, 'Audio-Magnetotelluric Study of La Réunion Island in Order to Estimate its Geothermal Potentiality', *Journal of Volcanology and Geothermal Research.* **20**, 311–332.

Benderitter, Y., Ballestracci, R., and Nougier, J.: 1983, 'Mise en évidence par sondages audiomagnétotelluriques d'une zone électriquement conductrice sous les plateaux basaltiques de iles Kerguelen', *Compt. Rend. Acad. Sci. Paris*, Série II, t. 296, 833–838.

Benderitter, Y.: 'Investigation audiomagnétotellurique et recherche d'une cible géothermique dans la région de Morne Rouge (Martinique)', *Rapport B.R.G.M.* **83**, SGN 725 IRG.

Bostick, F. X. Jr.: 1977, 'A Simple almost Exact Method of MT Analysis', in *Workshop on Electrical Methods in Geothermal Exploration*, Report of the Geothermal Workshop, University of Utah, Salt Lake City.

Bostick, F. X. Jr., Smith, H. W., and Boehl, J.: 1977, 'Audio-magnetotelluric Investigation on Kilauea Volcano and Kileuea Lava Lake', Final Report to Sandia laboratories; Electromagnetic Research Group, Department of Electrical Engineering, University of Texas, Austin.

Dupis, A.: 1971, 'Premières applications de la magnéto-tellurique à la prospection pétrolière, géologique ou minière de diverses régions métropolitaines', *Ann. di Geofis.* XXIV, n°2, 145–286.

Gamble, T. D., Goubau, W. M., Goldstein, N. E., and Clarke, J.: 1980, 'Referenced Magnetotellurics at Cerro Prieto', *Geothermics* **9**, 49–63.

Gamble, T. D., Goubau, W. M., Goldstein, N. E., Miracky, R., Stark, M., and Clarke, J.: 1981, 'Magnetotelluric Studies at Cerro Prieto', *Geothermics* **10**, 169–182.

Goldstein, N. E. and Mozley, E.: 1978, 'A Telluric – Magnetotelluric Survey at Mount Hood, Oregon – A Preliminary Study', Report No LBL-7050, Berkeley, Lawrence Berkeley Laboratory.

Goldstein, N. E., Mozley, E., and Wilt, M.: 1982, 'Interpretation of Shallow Electrical Features from Electromagnetic and Magnetotelluric Surveys at Mount Hood, Oregon', *Journ. Geophys. Res.* **87**, 2815–2828.

Grillot, L. R.: 1973, 'Regional Electrical Structure Beneath Iceland as Determined from Magnetotelluric Data', Ph.D. Thesis, Dept. of Geological Sciences, Brown University, 80, pp.

Hermance, J. F., Thayer, R. E., and Bjornsson, A.: 1976, 'The Telluric-Magnetotelluric Method in the Regional Assessment of Geothermal Potential', in *Proc. Second U.N. Symposium on the Development and Use Geothermal Resources* **2**, 1037–1048, Stock No. UG0-000-0000S-1, Supt. Documents, U.S. Government Printing Office, Washington, DC, 20402.

Hermance, J. F. and Pedersen, J.: 1980, 'The Deep Structure of the Rio Grande Rift: A Magnetotelluric Study', *J. Geophys. Res.* **85**, 3899–3912.

Hermance, J. F.: 1981, 'Crustal Genesis in Iceland: Geophysical Constraints on Crustal Thickening with Age', *Geophys. Res. Letters* **8**, 203–206.

Hermance, J. F.: 1982, 'Magnetotelluric and Geomagnetic Deep-Sounding Studies in Rifts and Adjacent Areas: Constraints on Physical Processes in the Crust and Upper Mantle', *Special Volume on Continental and Ocean Rifts* **8**, 169–192, International Commission on Geodynamics, AGU.

Hermance, J. F. and Colp, J. L.: 1982, 'Kilauea Iki Lava Lake: Geophysical Constraints on its Present (1980) Physical State', *Journ. of Volcanology and Geothermal Res.* **13**, 31–61.

Hoover, D. B., Frischknecht, F. C., and Tippens, C. L.: 1976, 'Audiomagnetotelluric Sounding as a Reconnaissance Exploration Technique in Long Valley, CA', *J. Geophys. Res.* **81**, 801–809.

Hoover, D. B., Brougham, G., and Clark, J.: 1976, 'Audiomagnetotelluric Data Log, Station Location Map, and Telluric profile Data for the Elko Hot Springs Known Geothermal Resource Area (KGRA)', Nevada: U.S.G.S.. Open-File Report 76–152.

Hoover, D. B., Long, C. L., and Senterfit, R.: 1978, 'Audiomagnetotelluric Investigations in Geothermal Areas', *Geophys.* **43**, 1501–1514.

Pedersen, J. and Hermance, J. F.: 1976, 'Towards Resolving the Absence or Presence of an Active Magma Chamber Under the Southern Rio Grande Rift Zone (abstract), *EOS* **57**, 1014.

Pedersen, J. and Hermance, J. F.: 1978, 'Evidence for Molten Material at Shallow to Intermediate Crustal Levels Beneath the Rio Grande Rift at Santa Fe (abstract)', *EOS* **59**, 390.

Pedersen, J.: 1980, 'A Magnetotelluric Investigation of the Deep Electrical Structure of the Rio Grande Rift and Adjacent Tectonic Provinces', Ph.D. thesis, Dept. Geol. Sci., Brown Univ., 29pp.

Pedersen, J. and Hermance, J. F.: 1981, 'The Deep Electrical Structure of the Colorado Plateau from Magnetotelluric Measurements', *J. Geophys. Res.* **86**, 1849–1857.

Pham Van Ngoc: 1976, 'Magnetotelluric Reconnaissance Survey in the Lilloet Valley', British Columbia, Open File Report, Earth Physics Branch, Energy Mines and Resources Canada, Ottawa, Canada.

Pham Van Ngoc: 1978, 'Magnetotelluric Prospecting in the Mount Meager Geothermal Region (British Columbia)', Open File Report, Earth Physics Branch, Energy Mines and Resources Canada, Ottawa, Canada.

Pham Van Ngoc: 1980, 'Magnetotelluric Survey of the Mount Meager Region and the Squamish Valley (British Columbia)'. Open File Report, Earth Physics Branch, Energy Mines and Resources Canada, Ottawa, Canada.

Slocum, W. M., Alvarez, L. J., and Hermance, J. F.: 1980, 'Combined Variation Gradiometry Measurements and Magnetotelluric Measurements in Iceland', *EOS* **61**, 225–226.

Stanley, W. D., Wahl, R. R., and Rosenbaum, J. G.: 1976, 'A Magnetotelluric Study of the Stillwater – Soda Lakes, Nevada Geothermal Area', USGS Open File Report No. 76–80.

Stanley, W. D., Boehl, J. E., Bostick, F. X., and Smith, H. W.: 1977, 'Geothermal Significance of Magnetotelluric Sounding in the Eastern Snake River Plain – Yellowstone Region', *J. Geophys. Res.* **82**, 2501–2514.

Stanley, W. D.: 1981, 'Magnetotelluric Survey of the Cascade Volcanoes Region', Pacific Northwest; Presented at 51st annual SEG meeting, Los Angeles.

Stanley, W. D.: 1982, 'A Regional Magnetotelluric Survey of the Cascade Mountains Region', U.S. Geological Survey Open File Report, 25 pp.

Stanley, W. D.: 1982, 'A Regional Magnetotelluric Survey of the Cascade Range Region', northwestern United States: U.S. Geological Survey Open File Report No. 82–126.

Thayer, R. E.: 1975, 'Telluric-Magnetotelluric Investigations of Regional Geothermal Processes in Iceland', Dissertation for the Degree of Doctor of Philosophy in the Dept. of Geol. Sciences at Brown University, 276 pp.

Thayer, R. E., Bjornsson, A., Alvarez, L., and Hermanence, J. F.: 1981, 'Magma Genesis and Crustal Spreading in the Northern Neovolcanic Zone of Iceland: Telluric-Magnetotelluric Constraints', *Geophys. J.* **65**, 423–442.

THE EISCAT-MAGNETOMETER CROSS

Operational Aspects – First Results

H. LÜHR, S. THÜREY, and N. KLÖCKER
Institut für Geophysik und Meteorologie, Technische Universität Braunschweig, Mendelssohnstrasse 3, D-3300 Braunschweig, F.R.G.

Abstract. A description of a new magnetometer network in northern Scandinavia is presented. The network, the 'EISCAT-Magnetometer Cross', is a cooperative project between Scandinavian and German research groups. It consists of seven continuously recording digital magnetometer stations. A special feature of the instruments is their almost maintainance free operation. The hardware has been developed at the Technical University of Braunschweig, F.R.G., and represents the state-of-the-art in digital magnetometer stations.

Operation began on 1 October, 1982 and the instruments functioned well during the first year. Some problems we had with the stability of the sensor mounts.

In spite of this we were able to gather a valuable set of magnetic field data.

1. Introduction

The characteristic features of the electric current systems associated with polar magnetic disturbances have been under investigation since the beginning of this century. The early studies were based on optical and magnetic observations made on ground (see e.g. Birkeland, 1908; Chapman, 1935). Later, in situ measurements using rocket- and satellite-borne instruments were added (reviewed by Potemra, 1979). New ground-based observation technique like the incoherent backscatter radar (e.g. Evans, 1969) have been employed recently. Nevertheless, even today ground-based magnetometer measurements provide the main data base for detailed analysis of the ionospheric-magnetospheric current systems (e.g. Hughes and Rostoker, 1979; Kamide *et al.*, 1982).

During the 'International Magnetospheric Study' (IMS) a number of ground-based magnetometer chains and arrays were operated around the world (summarized by Russell and Southwood, 1982). Many observations made with these magnetometers have been evaluated in conjunction with data from other facilities and gave new insights into the physics of auroral phenomena. After the IMS ended the magnetometer network in Scandinavia was removed. Data collection and preprocessing of both the Münster analog and the Braunschweig digital stations in this network made very heavy demands on manpower and travel (Küppers *et al.*, 1979; Maurer and Theile, 1978).

A new impetus to ground-based magnetic measurements in Northern Scandinavia was given with the advent of the 'European Incoherent Scatter' radar (EISCAT). It was decided that the fruitful cooperation of Scandinavian and German research groups built up during the IMS should be continued with a new ground-based magnetometer network to be operated in close relation to EISCAT. This new network, the 'EISCAT-Magnetometer Cross', is a joint enterprise of the Finnish Meteorological Institute

Geophysical Surveys **6** (1984) 305-315. 0046-5763/84/0063-0305$01.65.

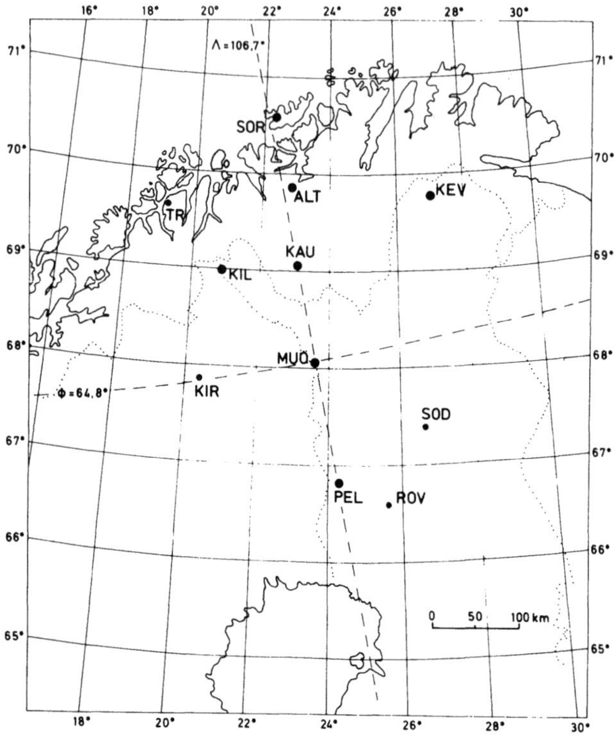

Fig. 1. Location of the seven stations of the EISCAT-magnetometer cross (fat dots).

TABLE I

Coordinates of the EISCAT-Magnetometer Stations

Station No.	Abbre-viation	Geographic		Corr. geomagnetic	
		Latitude	Longitude	Latitude	Longitude
010	SOR	70,54	22,22	67,30	107,90
020	ALT	69,86	22,96	66,62	107,81
030	KAU	69,02	23,05	65,78	107,16
040	MUO	68,01	23,53	64,73	106,66
050	PEL	66,90	24,08	63,56	106,02
060	KIL	69,05	20,70	65,98	105,45
070	KEV	69,76	27,01	66,24	110,60

(FMI), the Sodankylä Geophysical Observatory, the University of Tromsø, and the Technical University of Braunschweig. The distribution of responsibilities among these groups is as follows:

– The Braunschweig group: hardware procurement.

- Tromsø and Sodankylä: installation and the operation of the stations.
- Sodankylä: maintainance of the instruments.
- FMI, Helsinki: raw data processing and data tape production.

The project comprises seven temporary magnetometer stations. Five of them are aligned along a magnetic meridian near the Finnish-Swedish border. The remaining two stations are located approximately 100 km east and west of the chain (see Figure 1). The coordinates of the stations are given on Table I. The stations on the north-south chain began operation on 1 October, 1982 and since 1 July, 1983 the complete cross has been fully operational.

In the following sections we discuss the special features of the instruments and also give a resume of our experience we gained during 1 yr of operation.

2. Instrumentation

The instruments for the magnetometer stations have been specially developed with the specific intention of incorporating them in stations which require a minimum of maintainance. In practice this requires remote monitoring of the instrument operation. In order to realise the minimum-maintainance concept it is essential to have a number of precise and reliable components installed for use as references in an instrument self-check cycle. In our case we employed reference voltage sources and quartz oscillators temperature stabilized.

From experience with the IMS-magnetometers we identified three crucial areas for inprovement:

- a stable base line compensation network;
- a precise clock;
- a reliable data recording system.

Figure 2 describes the principle of operation in a block diagram. The heart of a station is the triaxial fluxgate magnetometer. Each axis of the magnetometer sensor is surrounded by a feedback coil. The current through this coil, representing the output signal of the magnetometer, is fed into the compensation network box. Here a switch selectable base line value can be subtracted from the input signal. To achieve the anticipated stability the whole box is thermostatially controlled. Figure 3 shows a simplified wiring diagram of the analog electronics. The output voltage which is a measure of the variation of a given magnetic component is fed to a data logger, where it is digitized in 12 bit words giving a resolution of 1 nT. The sampling rate is 3.2 sec^{-1} but the readings are averaged over a switch selectable period ranging from 2.5 to 40 sec.

A very stable quartz oscillator is used as the time base. This main oscillator not only controls the station clock but also the timing of the measuring cycle. To ensure a synchronous operation of all seven stations the complete control logic is reset at each change of the hour.

The scientific data, housekeeping data, and time information are commutated to a data frame starting with a synchronization pattern. This arrangement allows for serial data transmission via telephone lines or telemetry. In the normal mode the data are

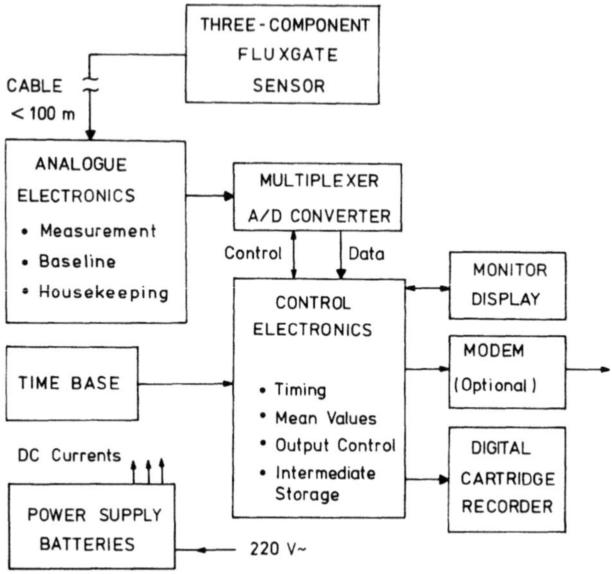

Fig. 2. Block diagram of a magnetometer station.

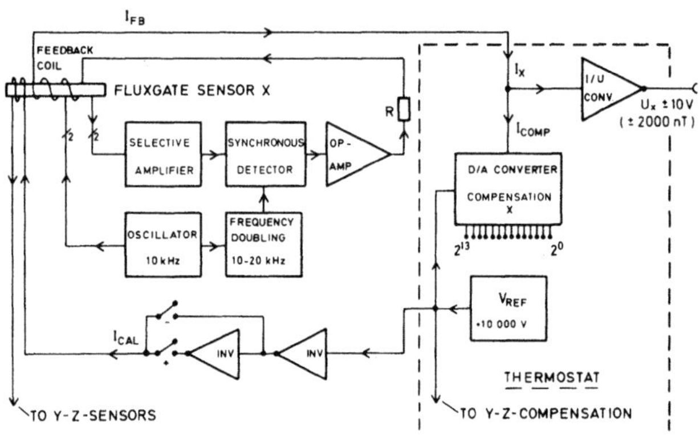

Fig. 3. Simplified wiring diagram of one magnetometer component. All critical components are housed in a temperature controlled box.

recorded on digital cartridge tapes, which are capable of storing 40 days' data at a typical data rate of three field vectors per minute. The power supply unit contains rechargeable buffer batteries which can operate the station for one day without AC power, and in addition contains a back-up battery to run the clock for a further 8 days. Once a day a self-check cycle is initiated, which include a check on the sensitivity of the magnetometer, the calibration factor and offset of the A/D converter, and the performance of the batteries.

The instrument parameters are summarized in Table II, and a photograph of a

TABLE II

Magnetometer Station Characteristics

Instrument	: Triaxial fluxgate magnetometer
Dynamic Range	: ±2000 nT
Resolution (12-bit ADC)	: 1 nT
Bandwidth	: 0.5 Hz
Sample Rate	: 3.2 / s
Averaging Period	: 2.5 sec through 40 sec (switch selectable)
Time accuracy	: ±1 sec
Housekeeping	: 13 channels
Instrument Status	: 12-bit flag word
Instrument Self-check	: once a day at 12 UT
Data Storage	: cartridge tape (DC 300 XL, Scotch)
Capacity	: 40 days (20 sec averages)
Power Supply	: 220 V ~, 24 V buffer batteries

complete magnetometer station can be seen on Figure 4. The option of real time data transmission via telephone line has not yet been used. This operational mode is planned only for special campaigns because of the rather high cost of dedicated lines particularly those communicating from one country to another.

In this mode, as soon as a connection is made interogation of instrument status is possible but there is no command capability foreseen. During the VIKING mission for example we intend to use this real time mode. Selected sites will then be accessible from

Fig. 4. A complete magnetometer station. The box on top contains the power supply unit. Beneath it we see the magnetometer electronics and the data logger, in front of which a data cartridge has been placed. On the right is the sensor assembly. The sensor tent is closed in normal operation. Overall dimensions can be estimated from the width of the 19 inch-racks.

the scientific center in the read-only mode. We plan to use both real-time and near-real-time transmission with an intermediate storage of several hours of data.

On-site recording is done continuously. Every 14 days the data cartridges are changed by the local residents and sent to the FMI in Helsinki. Here the data are preprocessed and written onto computer compatible tapes. The format of these tapes has been published by Sucksdorff (1981).

3. First Experiences

After one year of operation we can draw a first résumé. In general terms one can say the electronics worked well. We noticed no timing errors larger than 1 sec and the

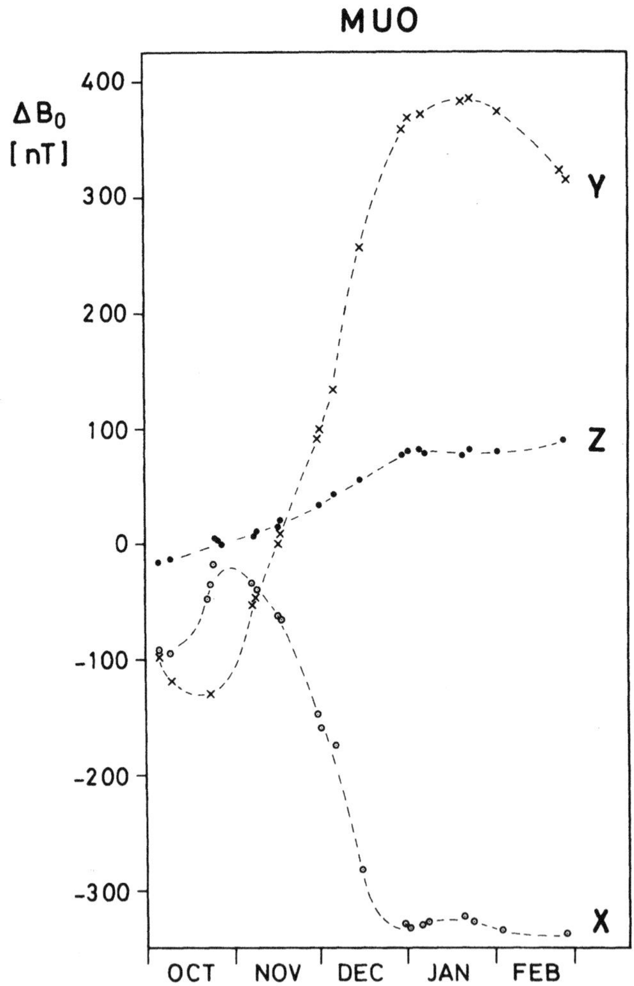

Fig. 5. Base line changes at Muonio. This is an example for pillar tilting caused by freezing of the ground.

uncertanty of the base line compensation network stayed within a range of $\pm 2\,\mathrm{nT}$. There were only two single component failures which were remedied by circuit board changes. A lot of trouble, however, was caused by unexpected large transients on the AC power lines at Sørøya. This resulted in a complete failure of the station after some months. Recently we have incorporated transient suppression devices and have not experienced any further interference.

The magnetometer sensors are mounted on concrete pillars approximately 50 m away from any building. At some sites the pillars did not satisfy stability requirements. Figure 5 shows the base line drift at Muonio due to the tilting of the pillar. Deeper pillars have been constructed in the meantime. We hope, that they will now be frost-proof.

An example of data from a sensor on a stable mount is presented on Figure 6. Figure 6(a) contains the differences between quiet day night values from Alta and Sodankylä.

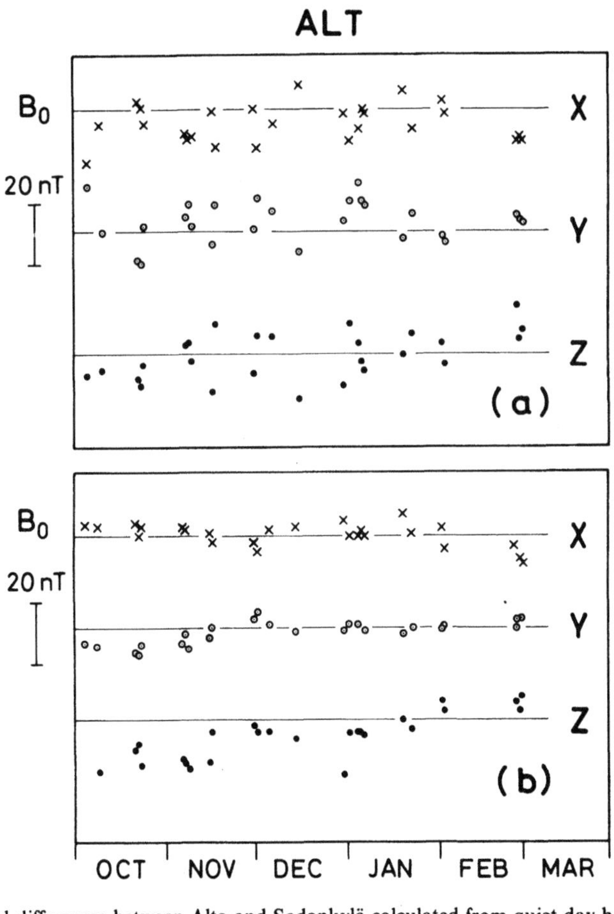

Fig. 6. Field differences between Alta and Sodankylä calculated from quiet day hourly mean values. (a) uncorrected, (b) corrected for sensor temperature changes.

The scatter of the data points can be reduced considerably by applying a temperature correction (see Figure 6(b)). The temperature coefficient of the sensors is of the order of 1 nT K^{-1}, and since the individual coefficients are known and have been stable for more than six months, the deviation can be corrected easily. Since a major objective of this project is the study of substorm related phenomena a stable base line is desirable but not mandatory. Our aim is to keep the changes below 5 nT per month.

The relative position of Scandinavia to the auroral oval calls for observation sites as far to the north as possible. At the land-sea interface, however, contributions to the magnetic field measurements from induction effects must be considered. The influence of induction has been studied in detail by Jones (1981) in this area. A coast effect can clearly be seen at the northern most station SOR. At ALT it is already smaller by a factor of four. Moreover this effect is much stronger during the hours around magnetic noon than around midnight. An explanation might be the inhomogeneity of the external field during substorm activity. More detailed investigations are necessary to find suitable correction algorithms.

4. Recent Scientific Results

During the last year a valuable data set has been gathered. One of the highlights, a pulsation event with very large amplitudes, will be presented here. The pulsations occurred on 2 November, 1982. Beginning around 17.00 UT with large horizontal disturbance vectors (see Figure 7). The onset was accompanied by mid-latitude Pi2 activity, the whole event lasting for about 40 min – the duration of the local substorm. The mean time period of the pulsations was 340 sec.

Latitudinal profiles of the magnetic deviations show a nearly two-dimensional structure of the ionospheric current sheets, with currents up to 4×10^5 A distributed over a width of only 20 km. The current bands drifted in a south-westerly direction with a velocity of 2.3 ± 0.2 km sec^{-1}. In the drifting frame of reference, these current systems are stable, the observed pulsations being produced as a result of their motion over the recording station. A more detailed description of this event is given by Lühr *et al.* (1984).

A comparison with riometer observations of the Finnish riometer chain is shown in Figure 8. The pairs of magnetometer and riometer stations have been selected such that both are aligned roughly to the electrojet. In the bottom panel the X'-component of the magnetic variations (X' perpendicular to the jet) is plotted, and in the middle we see the intensity of the ionospheric radio noise absorption. A good correlation between negative deviations of X' and absorption maxima can clearly by seen. To emphasize the minima of the X' component the autocorrelation function of X' has been added on the top.

These correlations can theoretically be explained in terms of kinetic Alfvén waves which were studied by Fejer and Kan (1969), Hasegawa (1977), Goertz and Boswell (1979), and Lysak and Dum (1983). The waves seem to be generated near the

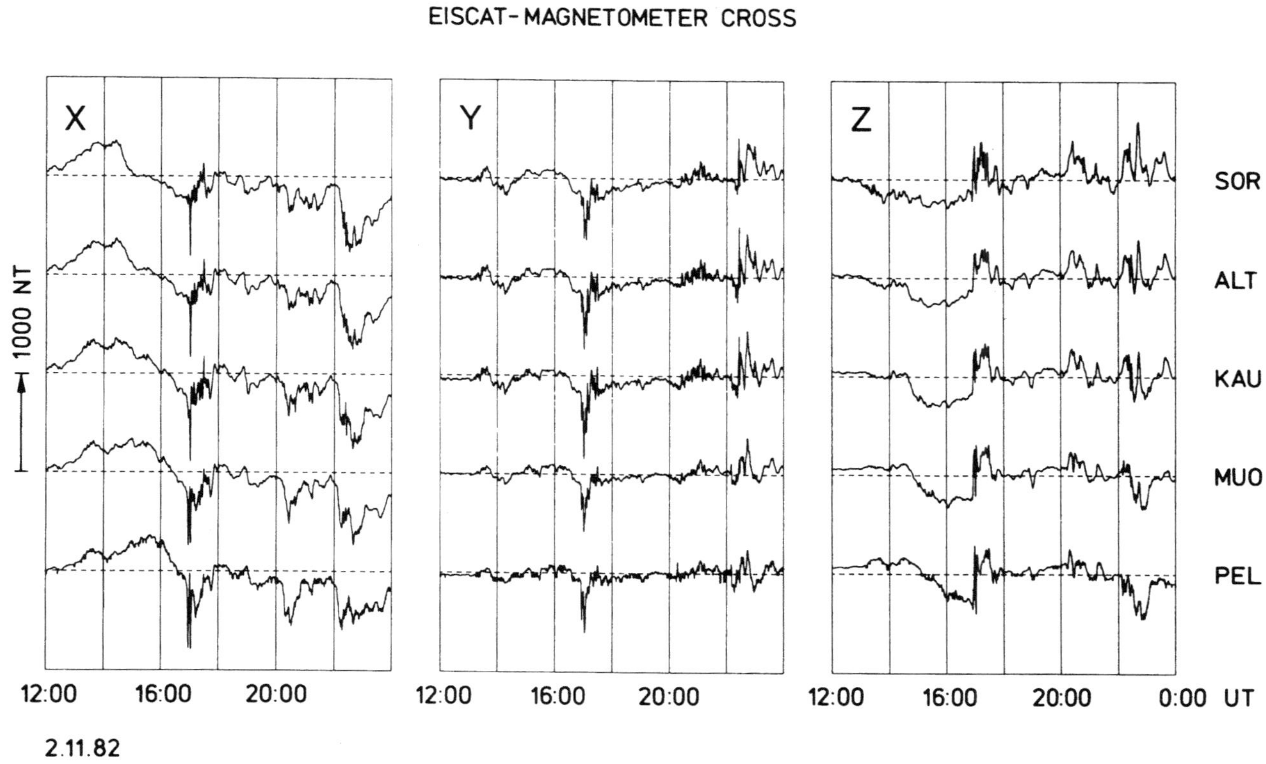

Fig. 7. Magnetic variations of the second half of 2 November, 1982 along a north-south chain in northern Scandinavia. The components X, Y, and Z are pointing towards geographic north, east, and downward, respectively.

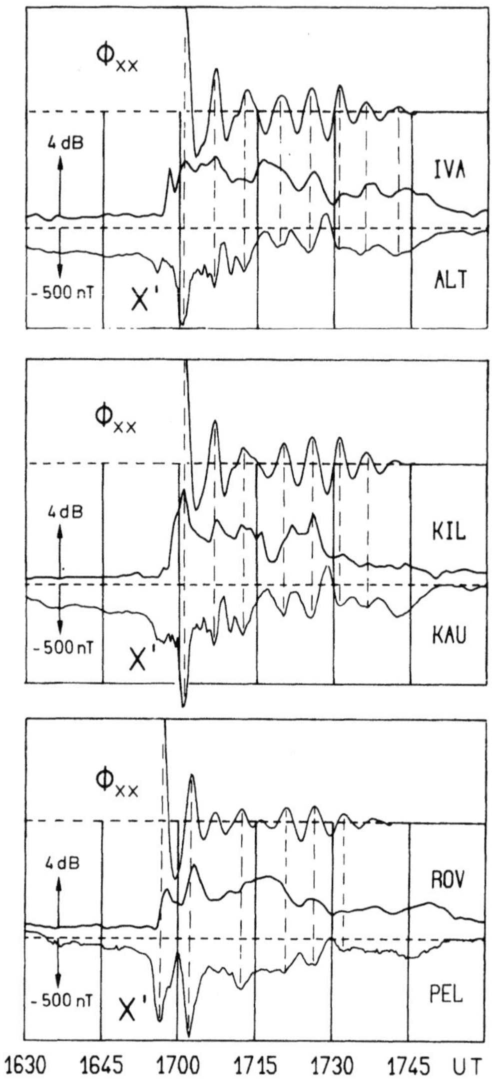

Fig. 8. Synoptical presentation of magnetometer and riometer recordings from related sites. The X'-component, displayed in the bottom panel, points to the north-east, perpendicular to the electrojet. The intensity of the radio noise absorption is plotted in the middle. There is a good coincidence between negative deflections in X' and absorption maxima. To emphazise the correlation the autocorrelation function of X' is added in the top panel (see vertical dashed lines). The pairs of sites are roughly aligned to the jet.

magnetospheric equator at the inner edge of the plasma sheet which drifted inwards during the active period.

The study of similar events will be one of the scientific objectives of our group, our aim being to establish general properties of the pulsations, and a more detailed physical explanation of them.

Acknowledgments

The EISCAT-Magnetometer Cross has been planned and installed in cooperation with the following institutes and persons: Finnish Meteorological Institute at Helsinki (Dr Ch. Sucksdorff), University of Münster (Dr W. Baumjohann (now MPI Garching), Dr H. Opgenoorth (now at Uppsala)), Sodankylä Geophysical Observatory (Dr E. Kataja, Mr P. Vilkki), University of Tromsø (Mr S. Berger), and Technical University of Braunschweig (Dr B. Theile (now with Dornier–System)). We are very grateful to all of them for their support and interest in the project.

The scientific output of the project is very much dependent on careful data evaluation. We are much obliged to Mr K. Lehto (Helsinki) and Mrs Ch. Willecke (Braunschweig) for their contributions to this aspect of our work. We are indebted to Dr P. Christiansen, MPI Garching for his editorial assistance.

The German part of this project is supported by grants from the Deutsche Forschungsgemeinschaft.

References

Birkeland, K.: 1908, 'The Norwegian Aurora-Polaris Expedition 1902–1903', *Christina* **1**.

Chapman, S.: 1935, 'The Electric Current-System of Magnetic Storms', *Terr. Magn. Atmos. Electr.* **40**, 349–370.

Evans, J. V.: 1969, 'Theory and Practice of Ionosphere Study by Thomson Scatter Radar', *Proc. IEEE* **57**, 496–528.

Fejer, J. A. and Kan, J. R.: 1969, 'A Guiding Center Vlasov Equation and Its Application to Alfvén Waves', *J. Plasma Phys.* **3**, 331–351.

Goertz, C. K. and Boswell, R. W.: 1979, 'Magnetosphere – Ionosphere Coupling', J. *Geophys. Res.* **84**, 7239–7246.

Hasegawa, A.: 1977, 'Kinetic Properties of Alfvén Waves', *Proc. Indian Acad. Sci.* **86**, 151–174.

Hughes, T. J. and Rostoker, G.: 1979, 'A Comprehensive Model Current System for High-Latitude Magnetic Activity – I. The steady State System', *Geophys. J. Roy. Astr. Soc.* **58**, 525–569.

Jones, A. G.: 1981, 'Geomagnetic Induction Studies in Scandinavia – II. Geomagnetic Depth Sounding, Induction Vectors and Coast-effect', *J. Geophys.* **50**, 23–36.

Kamide, Y., Ahn, B.-H., Akasofu, S.-J., Baumjohann, W., Friis-Christensen, E., Kroehl, H. W., Maurer, H., Richmond, A. D., Rostoker, G., Spiro, R. W., Walker, J. K., and Zaitzev, A. N.: 1982, 'Global Distribution of Ionospheric and Field-Aligned Currents During Substorms as Determined from Six IMS Meridian Chains of Magnetometers: Initial Results', *J. Geophys. Res.* **87**, 8228–8240.

Küppers, F., Untiedt, J., Baumjohann, W., Lange, K., and Jones, A. G.: 1979, 'A Two-Dimensional Magnetometer Array for Ground-Based Observations of Auroral Zone Electric Currents During the International Magnetospheric Study (IMS), *J. Geophys.* **46**, 429–450.

Lühr, H., Klöcker, N., and Thürey, S.: 1984, 'Ground-Based Observations of a Very Intense Substorm-Related Pulsation Event', accept for publ. *J. Geophys.*

Lysak, R. L. and Dum, C. T.: 'Dynamics of Magnetosphere-Ionosphere Coupling Including Turbulent Transport, *J. Geophys. Res.* **88**, 365–380.

Maurer, H. and Theile, B.: 1978, 'Parameters of the Auroral Electrojet from Magnetic Variations along a Meridian', *J. Geophys.* **44**, 415–426.

Potemra, T. A.: 1979, 'Current Systems in the Earth's Magnetosphere', *Rev. Geophys. Space Phys.* **17**, 640–656.

Russell, C. T. and Southwood, D. J.: 1982, *The IMS Source Book – Guide to the International Magnetospheric Study Data Analysis*, AGU, Washington, D.C.

Sucksdorff, C.: 1981, 'Tape Formats for Geomagnetic Data', *IAGA News*, pp. 112–113.

AN INTELLIGENT DIGITAL MAGNETIC RECORDING SYSTEM (DIMARS)

L. HEGYMEGI
Eotvos Lorand, Geophysical Institute of Hungary, Budapest
and
L. DRIMUSZ
Technical University, Budapest, Hungary

Abstract. The system consists of three magnetic variometers with electrical output signals for the *H*, *D*, and *Z* components, a proton magnetometer for the total field, a data acquisition system, two cassette recorders, an analogue chart recorder, a calendar clock, and a telex. The analogue and digital input data are sampled at 10 sec intervals, averaged over a minute, and recorded on a magnetic tape cassette. The computed hourly and daily mean values and the daily minimum and maximum values are recorded together with the occurrence times on a second cassette, at the operator's request. Rapid run recording of the 10 sec values of each measured component is initiated automatically during magnetic storms.

DIMARS incorporates several test functions as ROM test, A/D converter test, recorder test, telexline test and checking the memory content. The test functions can be activated by the equipment keyboard, or via the telex. The telex can also be operated as an output printer.

1. Introduction

The work of a conventional magnetic observatory is labour intensive. At most observatories magnetic indices and hourly mean values are produced in whole, or in part, by handscaling methods. Automatic digitising equipments, which can elaborate the photorecords, are available but they are rather expensive.

Since the handling of magnetic data requires a computer it is very convenient if the observatory data are in a machine readable form. For this reason digital recording systems are developed to record the data directly. The digital data are subsequently processed by computers to produce the calculated tabular values for the yearbooks.

Our measuring system, constructed for the Tihany Observatory, differs from other digital stations. It computes or memorises all the data normally published in an observatory yearbook. In addition to this it records on cassette the minute mean values of the magnetic elements and the total field. It also provides a facility for rapid run recording during magnetic storms.

2. System description

The block diagram of the recording system is shown in Figure 1. Conventional quartz magnetometers are used as signal sources. Each magnetometer is equipped with a feedback coil system. The light beam reflected from the magnet/mirror is converted into an electrical signal by a photo-sensor. When the field varies a current proportional to the detected signal is fed back to the coil to create a compensation field which opposes

Geophysical Surveys **6** (1984) 317-321. 0046-5763/84/0063-0317$00.75.

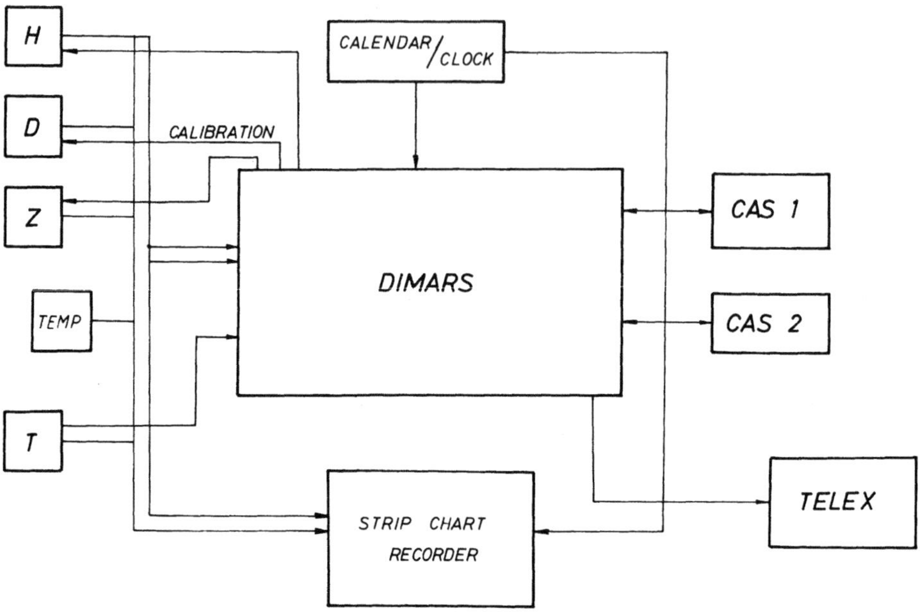

Fig. 1. The block diagram of the digital magnetic recording station of the Tihany Observatory.

the deflection of the magnet. The magnitude of the coil current, which is proportional to the magnetic field variation, is measured by the data logger.

Any type of variometer with an electrical output can be used as a signal detector. The measurement accuracy of the system basically depends on this part of the equipment.

A proton magnetometer is also connected into the DIMARS system. This instrument measures the absolute value of the total field vector and provides a digital output. At Tihany the proton magnetometer is located some distance (about 50 m) from the data logger. Serial data transmission is effected using a 3 core inter-connecting cable.

Date and time information for the system is provided by a quartz calendar clock. This is connected to the system by a serial link in the same way as the proton magnetometer. The clock also supplies the time markers for the analogue recorder.

Two Memodyne type 143 cassette recorders are used for recording the data. Minute mean values are recorded on cassette I. Cassette II is used to record rapid run data and the month's hourly and daily mean values.

All the important functional characteristics of the measuring system can be controlled or programmed by the telex. It is also possible to transfer the measured or calculated data to the telex.

Because the variometers are sensitive to variations in temperature the measuring room is thermostatically controlled. The data logger carries out this control function and continuously monitors the amplitude of the signal input to each channel and the status of the tape recorders. In the event of any malfunction being detected a warning appears on the telex.

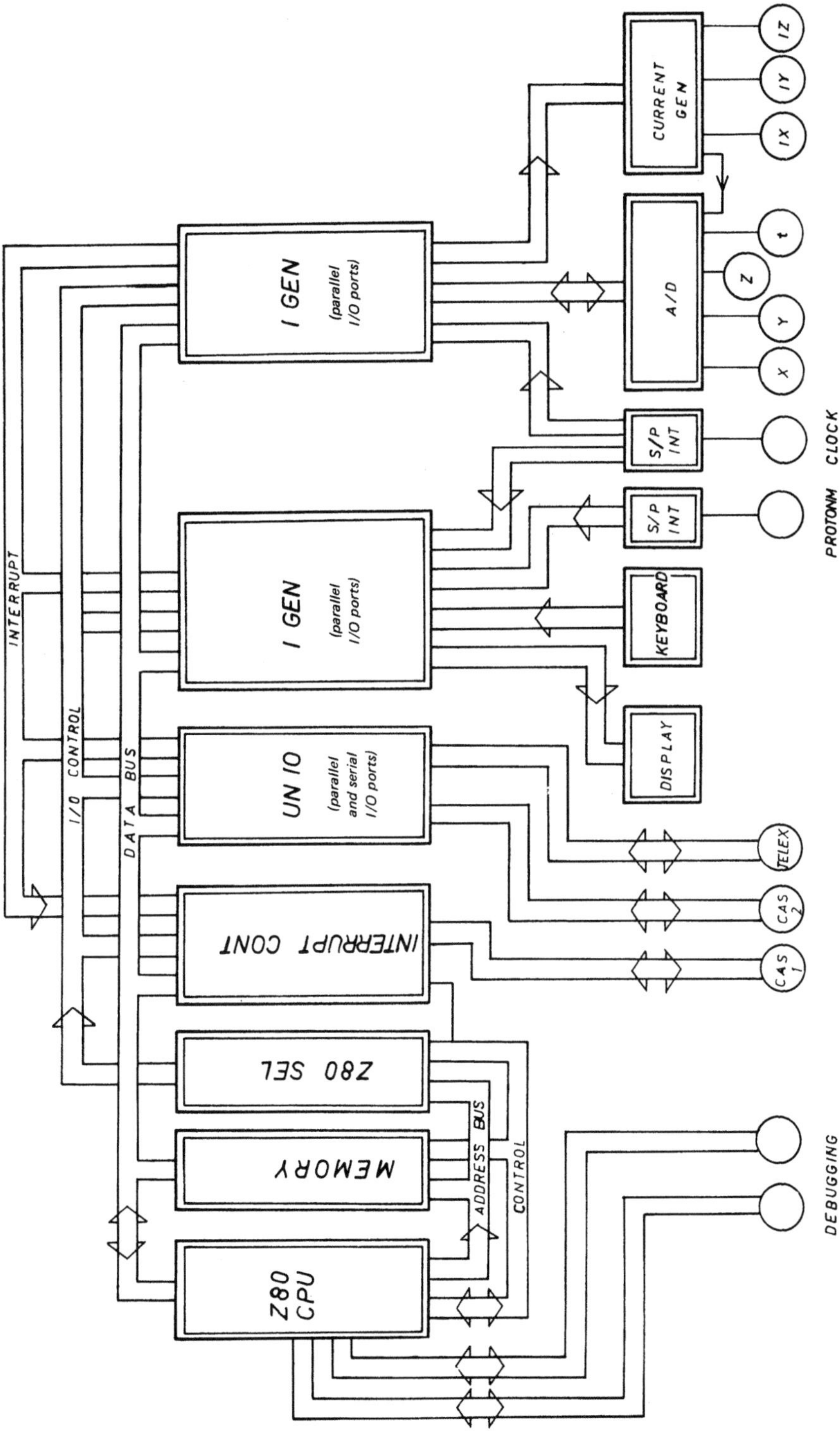

Fig. 2. The block diagram of the intelligent magnetic data acquisition system.

A block diagram of the data logger is shown in Figure 2. It is a microprocessor controlled analogue data logger, equipped with suitable digital peripheral devices.

The control unit is a Z80 microprocessor with a maximum memory capacity of 24 Kbyte. This includes ROM-s and RAM-s. Their relation depends on the system programme and memory size. The block labelled Z80SEL provides the selector addresses for the input/output (I/O) ports. The display, keyboard, A/D converter and input multiplexer, the serial and parallel inputs/outputs, and the current generator, which provides calibration currents for the variometers, are connected to the I/O ports.

The main technical characteristics of the system are:

- number of input channels: 5 (expandable to 8), differential, floating
- input resistance: 100 Ω
- input amplitude: 0.8192 V maximum in 3 ranges with automatic ranging and polarity
- maximum common mode input noise: 4 V
- digital input: serial, synchronous, TTL, 20 bits
- clock input: serial, synchronous, TTL, 31 bits
- calibration outputs: 3 floating, programmable current outputs, 0 - ±5, ±10, ±15 mA. Maximum load resistance 100 Ω
- recorder outputs: 8 bit, parallel, TTL and 6 bit command lines
- telex: serial, asynchronous, duplex, current loop, 50 baud.

3. System Operation

The A/D converter measures the voltage input to each channel at 2 sec intervals. The measured values are memorised every 10 sec and mean values are calculated every minute. Minute mean values are recorded on cassette recorder I and stored in memory for further calculations. Hourly mean values are calculated from the minute values. The maximum and minimum values for each channel and their times of occurrence are continuously monitored and stored for each day.

At the end of the day the daily mean values are calculated. All the mean values for the last 3 days can be interrogated by telex. In the 24 hr period following the end of a month the hourly and daily mean values for the preceding month can be recorded on cassette recorder II without any loss of data.

During the periods when the mean values are being written onto tape, or the cassette is changed, no data loss is incurred because the measured data are stored in the memory and recorded on the cassette recorders after the end of the operation.

The format of the records on cassette recorder I is:

- At midnight the data block contains: the calibration values of all the variometers; the baseline values; the temperature of the measuring room.
- Each 10 min duration data block contains: the year, month, date, and time; 10 min mean values of H, D, Z, and T.

Cassette Recorder II records the rapid run data and the mean value tables. Rapid run recording starts automatically when the variation in the H component fulfills the conditions:

$$\Delta H > A/10' \quad \text{or} \quad \Delta H > B/1^h.$$

Where A and B are constants pre-programmed by the operator and ΔH is the actual change in the H component. The rapid run registration can be terminated by the operator.

Continuous monitoring of the system operation is possible by means of the display. The working mode of the display can be selected on the keyboard. All measured and calculated data may be viewed on the display as can the self test results.

4. Data Processing

There are two possibilities:

In order to print the mean value tables for the yearbooks we can simply connect a printer to the cassette recorder. If however the baseline changes with time, baseline corrections must be applied before preparing the final tables. The temperature stability of the system is not sufficiently known at the present time but we are hopeful that baseline variations can be eliminated completely by temperature stabilisation of the measuring room.

The alternative method involves the transcription of data from cassette to 1/2″ magnetic tape for subsequent processing of the minute mean values and rapid run data. At present transcription and further processing are carried out on a PDP 11/40 computer.

ON THE REDUCTION OF TIME VARIATIONS FOR GEOMAGNETIC REPEAT STATION MEASUREMENTS

GÜNTER SCHULZ
Deutsches Hydrographisches Institut, Erdmagnetisches Observatorium, Am Olymp 13, D-2177 Wingst, F.R.G.

and

MARTIN BEBLO
Geophysikalisches Observatorium, Ludwigshöhe 8, D-8080 Fürstenfeldbruck, F.R.G.

Abstract. For most previous geomagnetic surveys, the time variations of the observatory closest to the site of observation have been used for reduction. The accuracy of repeat station measurements depends upon the distance between the station and the observatory. Nowadays, the requirements for the accuracy of repeat station measurements are much higher, especially with respect to local secular variation anomalies. On the other hand, transportable variograph stations are available, by means of which time variations can be recorded closer to the station site.
The reduction steps for this method are:
(1) Reduction of measured values to the base-line of the nearby variograph station;
(2) Computation of the nearest quiet night value using the baseline of the variograph;
(3) Computation of the difference between this value and the corresponding one at the observatory.
The problems are:
- the stability of the base-line values of the variographs;
- the question as to what degree a quiet night value represents the normal value at that time (e.g. influence of a D_{st} field and/or other fields).
A repeat station survey of the Federal Republic of Germany was carried out in 1982. During the same time, eleven stations recorded the time variations in that area.
We discuss the aforementioned aspects by way of an example taken from this survey.

1. Introduction

It is not the purpose of this paper to present the complete result of a geomagnetic survey, but to point out some of the difficulties with which one has to cope if one intends to include variations at temporary stations in the survey's evaluation.

As a first approximation, two field measurements carried out at different epochs are sufficient for measuring the secular variation (SV). Basic in the calculation of the SV from the measured values are the normal values of an observatory in the vicinity. The crucial question is: how exactly can one approximate a normal value at a secular point? If the field of the variations over the area to be surveyed were homogeneous, the certainty of the result would only be limited by the uncertainty of the instrument used.

Submitted to the XVIII General Assembly of IUGG, 15 to 27 August 1983, Hamburg.

Geophysical Surveys **6** (1984) 323-332. 0046-5763/84/0063-0323$01.50.

However, in principle, the field of the variations is inhomogeneous in respect to space and time. Severely simplified, one can say:

- the local and the regional differences of the short-term variations with periods of about (1 h) are caused by induced internal parts, which are, in principle, period-dependent but time-independent (except for phase shifts within the effects themselves);
- on the contrary, the sources of the inhomogeneous parts of the S_q variations are mainly located external to the Earth.

2. Variations

Before going into details, we give first a description of the geographical positions of some observatories and variometer stations. Figure 1 shows the area of the Federal Republic of Germany (F.R.G.) and the nearby observatories Niemegk (NGK), Budkov (BDV), Fürstenfeldbruck (FUR), Dourbes (DOU), Witteveen (WIT), and Wingst (WNG). In addition, the permanent recording stations Braunschweig (Brs), Göttingen (Gtt), and Münster (Mst) are shown. Of the temporary stations, in this connection, only Mellrichstadt (Mel) is of interest. Also shown are the secular points Haimbach (hmb) and Oberbimbach (obb). The distance between FUR and WNG is some 600 km. The aforementioned two secular points and Mel bisect this distance approximately.

Now let us consider the Sq variations, again. Figure 2 shows the differences of Sq variations in the sense FUR-WNG, Mel-WNG, and FUR-Mel for April 1982. A purely latitudinal dependence is to be seen, because the longitudinal differences between these stations can be neglected.

The horizontal component H shows the greatest effects. Therefore, the following statements are restricted to this component only. Beside the expected smooth shapes of the curves during the night hours and the decrease of the amplitudes from North to South, phase shifts can be recognized. Due to the phase shift, for secular points located between these stations, one is only permitted – to a certain degree – to reduce by interpolating the amplitudes linearly. In the early hours of the morning, one would obtain false results. For instance, in the worst case, at 07.30 UT both Mel-FUR and Mel-WNG become -15 nT, if one refers to the daily mean value indicated by the horizontal line. It follows that the result would be erroneous by the same amount, if one reduces using data from either WNG or FUR. On the other hand, the result would not depend upon the variograph station, if both differences vanish. This case almost happens at 14.30 UT. The method of interpolation has already been applied by Voppel and Wienert (1974), when evaluating the last-but-one survey of the F.R.G. in 1965.0.

The latitudial dependence of the S_q field shows a seasonal dependence, as can be recognized by comparing the traces of Figure 2 with the curve of *H* for FUR-WNG for October 1982 in Figure 3.

The longitudinal behaviour of the S_q field is well known. However, it is only useful to take it into account, if the area shows a large extension in the eastwest direction. In any

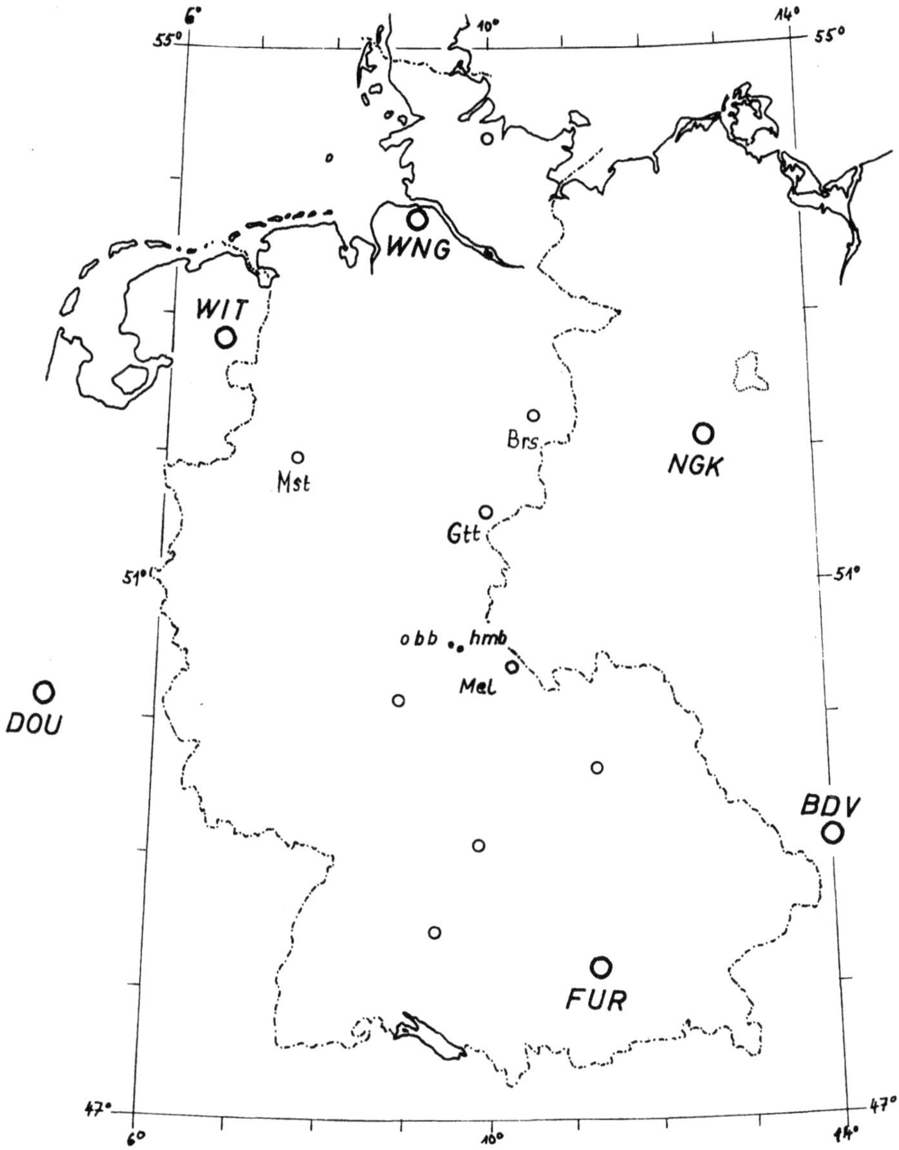

Fig. 1. The area of the Federal Republic of Germany; (○) neighbouring magnetic observatories; (∘) permanent recording stations and temporary stations; (·) secular points Haimbach and Oberbimbach

case, for an area of the size of the F.R.G. the effect would amount to 5 nT at most. However, the reduction is not simple, because the Sq field has superimposed on it the effects of particle radiation and daily inequalities, which are difficult to separate (Voppel and Wienert, 1974).

In conclusion, one can say: the inhomogeneous parts of the variation field can be neglected only if the variations themselves become small. For S_q, to a first approximation, this occurs during the night hours from 00 to 03 of local time. Bay type disturbances must not occur at all.

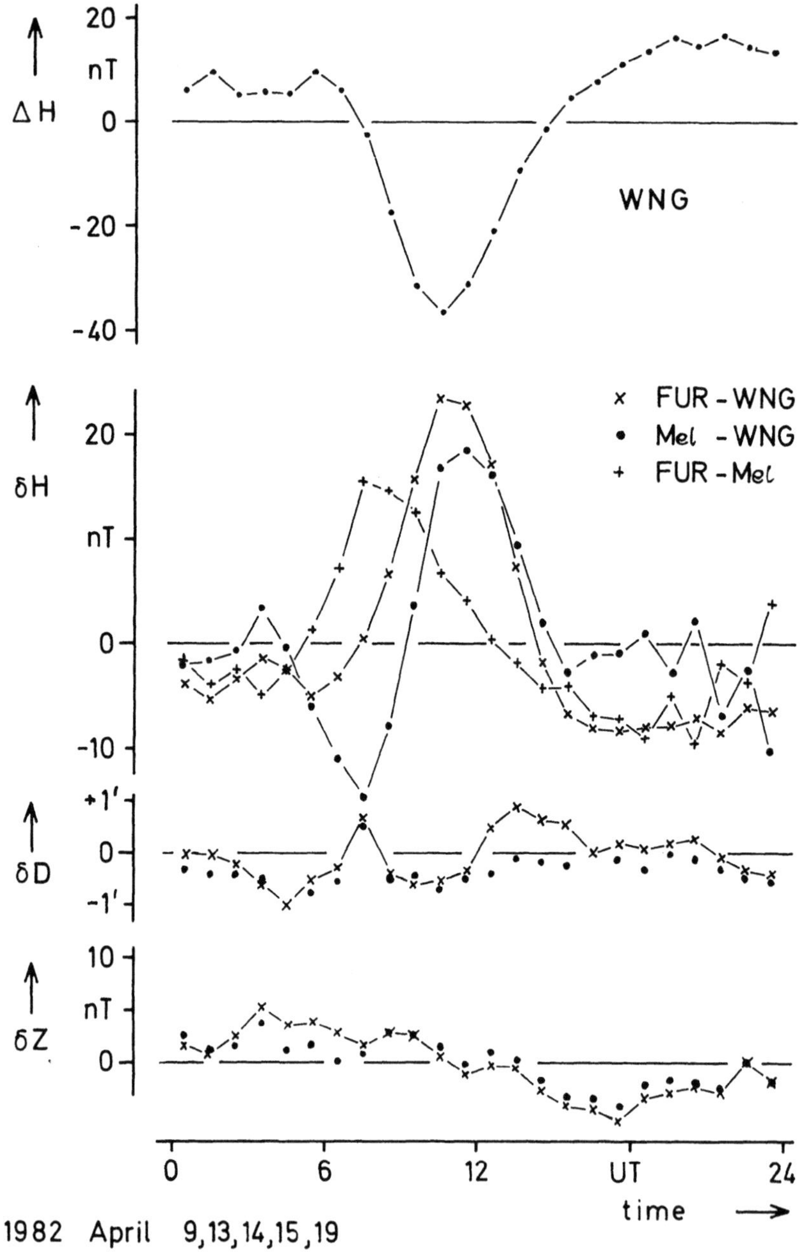

Fig. 2. S_q variations and their differences in April 1982.

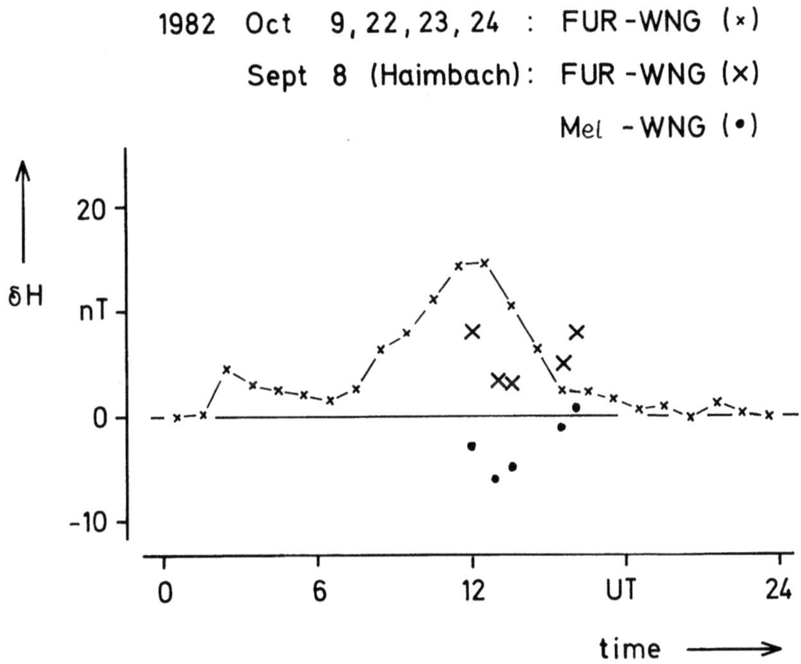

Fig. 3. Differences of S_q variations of H in October 1982: enlarged symbols: differences of variations on 8 September 1982 (see Figure 5).

If one selects quiet night hours in accordance with these aspects, the inhomogeneity over an area of the size of the F.R.G. amounts to not more than ± 2 nT. This has been shown by a comparison of momentary values between European observatories, which has been running for nearly three decades.
How far influences of the equatorial ring current (D_{st}) play any part has to be examined for each individual case. It is necessary that one can identify and separate the external influence which led to a shift of the quiet night hours' levels. In the case in question, the investigation reduces to a judgement of the local difference between FUR and WNG at the quiet night hours with regard to a long-term average (called 'normal') of H.

3. Instruments

The modern measuring technique has reached a state which allows one to achieve results with a residual uncertainty of ± 2 nT, i.e. the same order of magnitude as inhomogeneities during quiet night hours.

Whereas, in 1935 (Bock et al., 1948) the inclination I was measured with an earth inductor, in 1965, Voppel and Wienert (1974) calculated I from H, measured by means of quartz horizontal magnetometers (QHM and HTM) and from the total intensity F, measured by means of a proton magnetometer (PRM) (Table I). In the late seventies, the development of the probe theodolite – with a fluxgate as null indicator – was

TABLE I

Instruments used during the surveys 1935.0, 1965.0, and 1982.0.

Epoch	*D* *H*	*Z*	*I*	*F*
1935.0	travel theodolite	field balance	earth-inductor	–
1965.0	QHM/HTM	(torsion magnetometer)	–	proton magnetometer
1982.0	probe theodolite (QHM)	–	probe theodolite	proton magnetometer

perfected. This instrument permits absolute measurements of I (and, of course, of the declination D) with a high precision.

However, in the early sixties, the PRM brought about the most essential improvements of the measuring technique. Therefore, the SV from 1965 to 1982 will be based upon results achieved from instrumental sets which can be considered comparable with regard to accuracy. This applies particularly to Z and F, the SVs of which, with regard to the period from 1935 to 1965, are obviously falsified by the errors of the survey of 1935 (Voppel and Wienert, 1974). Therefore, if one uses the modern instruments, the accuracy of the results depends largely upon the inhomogeneity of the variations.

4. Procedure of Evaluation

The improved modern measuring technique, the availability of several transportable variographs, and the findings gained during the 1965.0 survey, have been decisive for the development of the concept of the 1982.0 repeat survey, which is the subject of this paper:

(1) Reduction of measured values to the base-line of a nearby variograph station.
For this purpose, the existing network of recording stations has been segmented by six variographs. The variographs are fibre suspended magnet systems.

(2) Computation of the nearest quiet night value, using the base-line of the variograph (first step of the changeover from the variograph to the observatory).
A sufficiently high stability of the base-lines during the operating period is a prerequisite of this method. Figure 4 shows the behaviour of the base-lines of the variograph of the station Mel. The relative base-lines have been calculated by means of comparing the momentary values (02 UT) with the neighbouring observatories NGK, FUR, DOU, and WNG (circles) and FUR and WNG alone (dots), respectively. Whilst D shows a satisfactory stability of its base-line (± 1 nT), the scatter of the other components (H and Z) is about ± 3 nT. This fact is particularly noticeable for the

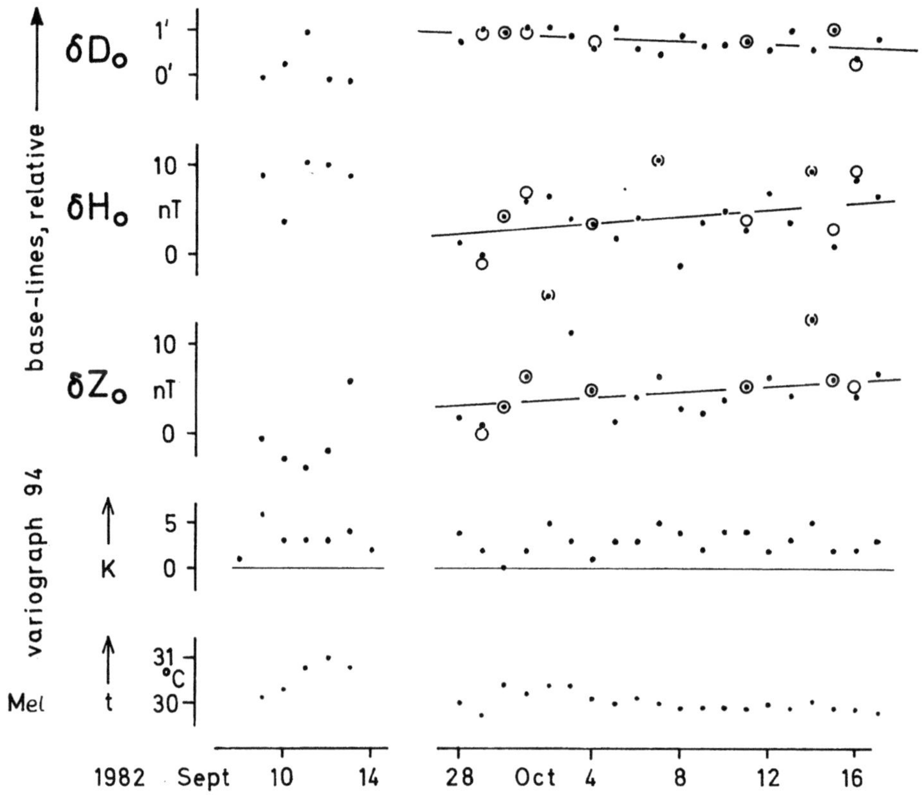

Fig. 4. Upper part: relative base-lines of the variograph 94 (temporary station Mellrichstadt); lower part: K indices of Wingst Observatory as a measure of corpuscular disturbances; temperature inside the variograph.

values from 29 September to 1 October, 1982. Non-compensated parts of the temperature coefficients may be the reason for scatter of the base-lines. The variation of the temperature is shown in the lowermost trace. Deviations of up to 0.5 °C can be seen during this period.

(3) Computation of the difference between the variograph's quiet night value and the corresponding one at the observatory (second step of the changeover from the variograph station to the observatory).

After the aforementioned remarks, this value approximates the difference between the station and the reference observatory quite well, if one disregards variations.

5. Examples

The measurements have been carried out by the observatories FUR and WNG during the period from March to October 1982 and have been sponsored by the Deutsche

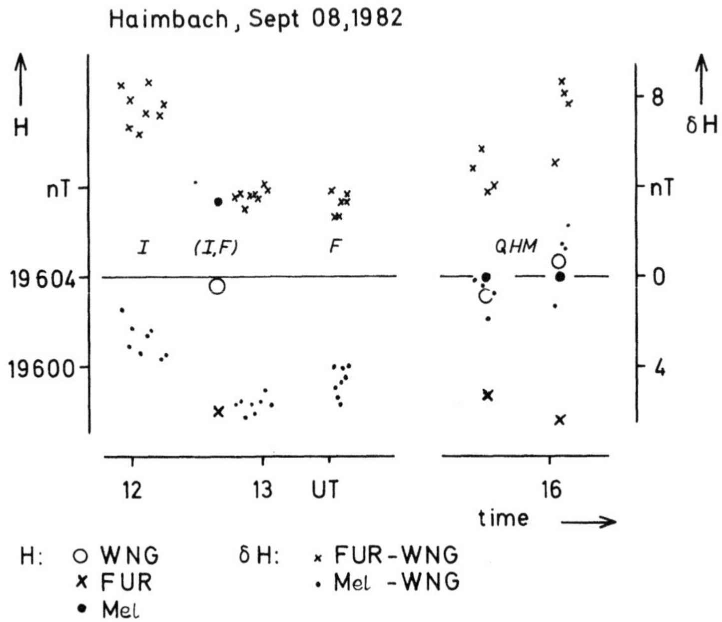

Fig. 5. Results of the measurement of H at the secular point Haimbach on 8 September, 1982, as a function of UT; (×)(•) (right scale δH): differences of the variations in the sense FUR-WNG and Mel-WNG, respectively; (o)(×)(●) (left scale): quiet night hour values of H at the station site reduced according to WNG, FUR, and Mel, respectively.

Forschungsgemeinschaft.

Figure 5 shows the result of H at the secular point Haimbach on 8 September. D and I have been measured by means of a probe theodolite, F by means of a PRM and, in addition, D and H by means of two QHMs. One can see the variations of FUR (crosses) and Mel (dots), referred to the variations of WNG (horizontal line), as a function of UT, as far as the variations have been necessary for the reduction to the base-line. The left-hand ordinate shows the component at the time of the changeover during quiet night hours. The circles show H, if one reduces according to WNG, the crosses, if one reduces according to FUR. The mutual difference of some 5 nT has to be traced back not only to S_q variations, as one can see from the enlarged symbols in Figure 3. Obviously, short periods of corpuscular origin contribute to the difference. The magnetic disturbance was measured as $K = 3$, at that time.

Because of the general behaviour of the S_q variations, the values reduced according to Mel are expected to lie between the values reduced according to FUR and WNG. That should be particularly true at noon. However, in the case in question, the result deviates largely from the value expected. On the other hand, for the QHM measurements which followed shortly afterwards, the reduction according to Mel gives an approach of the two results to each other. Therefore, one can come to the conclusion that the measurement carried out with the probe theodolite is erroneous. It is more likely that a motion of the variograph's base-line during the measuring time caused this difference, because the variograph worked only intermittently. The variograph only became ready for operation just before the measurement started. The behaviour of the base-line in Fig. 4 shows that the instrument ceased operation again after five days; namely, on 14 September, 1982.

For a reduction via the variograph, the measurement at the secular point Oberbimbach in the immediate vicinity of Haimbach on 5 October offers more favourable conditions. The changeover can be carried out during the night from 4 to 5 October.

At the time of changeover, the difference of H between the observatories in the sense WNG-FUR amounts to −2706 nT; thus lying some 1 nT above the mean value of −2707 nT in September/October. Therefore, the difference can be denoted as 'normal', i.e. probably free from external fields. On the contrary, the equivalent value at Haimbach is −2701 nT – an indication of the influence of D_{st} (or D_{st}-type currents) the influence of which decreases from South to North.

At this time, the negligible degree of corpuscular disturbance is symbolized by $K = 1$. The result can be seen from the following Table:

reduction according to:		Mel	FUR	WNG
component	UT			
D	17.50	−1°19.8′	−1°20.0′	−1°20.4′
I	18.20	65°53.9′	65°54.1′	65°54.0′
F	18.50	47936 nT	47936 nT	47935 nT

Appreciable differences appear only in *D* and *I*. The difference in D is presumably real, as opposed to that in *I*, which lies within the scatter of the variograph's base-line for *H*.

6. Conclusion

The examples discussed can only indicate how the method works, under which conditions it fails or how it can be improved by modifications:

(a) More stable *Z* base-lines of the variographs are likely to be achieved if one temperature compensates the variograph's systems more carefully before their employment. Contrary to *Z*, the *H* compensation does not depend on the component itself. Therefore, it has been preadjusted by the manufacturer and cannot be changed. In this case, presumably, a temperature stability of ±0.5 °C does not suffice.

(b) The *Z* base-line has been upset frequently by artificial disturbances (motor cars), making the tracing of the tracks very laborious. Therefore, installation at predominantly undisturbed places should be a prerequisite in the near future.

(c) The operation of a variograph demands regular maintainance, so as to avoid interruptions of the recording and, thereby, loss of valuable information.

The intention of the method described was to diminish not only the influence of solar electromagnetic radiation, but also the influence of particle radiation. Nevertheless, because of the aforementioned instrumental difficulties, it appears advisable to carry out the coming repeat survey during a sunspot minimum, as was the case in 1965. On average, over the periods of measurements, the corpuscular disturbances amounted to $a_k = 22$. For 1965, the equivalent value was only $a_k = 6$.

Acknowledgement

We wish to thank Prof. D. Voppel who suggested publishing this paper and provided many valuable critical comments.

References

Bock, R., Burmeister, F., and Errulat, F.: 1948, *Magnetische Reichsvermessung 1935*.0, Teil I (Tabellen), Abh. Geophys. Inst. Potsdam, No. 6.

Voppel, D. and Wienert, K.: 1974, 'Die geomagnetische Vermessung der Bundesrepublik Deutschland, Epoche 1965.0', *Dt. hydrogr.Z.* **27**, No. 2.

This group contains contributions from magnetic observatory operators on the methods by which observatory data can be recorded digitally. Digital data is required by all scientific and commercial users, and reluctance by the observatories to cease using classical methods has been based on the fear of losing the necessary high standard of recording and understanding of the geomagnetic field and its variations. Economics and technology combine to remove that fear and the collections shows the form of magnetic observatories in the next generation.

The paper about the Antarctic observatory underlines, perhaps in an exaggerated way, the difficulties of making precise measurements of the Earth's magnetic field. No magnetic observatory is perfectly situated. There are many reasons why its measurements may be systematically or transiently in error, e.g. local crustal magnetism or electrical conductivity or man made interference. While field models and hence charts depend on ground observatories and repeat stations, the best possible global coverage is necessary and continued scrutiny of methods and measurements is essential.

ONE YEAR EXPERIENCE WITH THE CMVS 2 MAGNETIC VARIATION STATION

J. PODSKLAN and I. KOVAC
Geophysical Institute, Slovak Academy of Sciences, 94701 Geomagnetic Observatory, Hurbanovo, Czechoslovakia

Abstract. A year's operation of a digitally recording variometer (CMVS 2) employing a Bobrov magnetometer is described. The equipment provides hourly mean values of the geomagnetic component on paper tape. It has a range of outputs including analogue recording of the field variations which can be measured in the range 0–1999.9 nT with a sensitivity of 0.1 nT. After a running in period of some months the standard deviation of baseline values is in the range 1.5–2 nT with a drift rate of about 2 nT per month.

1. Introduction

The purpose of this report is to summarise the results obtained in the course of one year of operating the CMVS 2 digital magnetic variation station. The station is one of the first three produced at the SKB FP of the U.S.S.R. Academy of Sciences. Employees of the organisation installed the equipment at the Geomagnetic Observatory in Hurbanovo in March, 1982.

2. Brief Description of the Equipment

The CMVS 2 consists of three principal parts: the assembly containing the D, H, and Z variometers, the variometer electronics, and the measurement unit. The sensitive part of the equipment consists of Bobrov-type variometers with photoelectric conversion and negative feedback. This pattern of variometer was chosen because of its excellent time and temperature stability.

The CMVS 2 provides instantaneous values of the field components together with information from an external data channel, and hourly mean field values which are output, via an interface, to a tape punch. The facility for recording real time variations of the geomagnetic field components on analogue recorders is also available. The options for connecting external equipment to the CMVS 2 are very wide (see Figure 1). The measurement range of the geomagnetic field variations is 0–1999.9 nT. The resolution is 0.1 nT, and the frequency range 0–5 Hz.

3. Recording Conditions

The CMVS 2 is installed in two buildings. The variometers and their control electronics are installed in the variometer house. The measurement unit is in another building 80 m distant.

During the period of the investigation the variometer house temperature was

Geophysical Surveys **6** (1984) 335-338. 0046-5763/84/0064-0335$00.60.

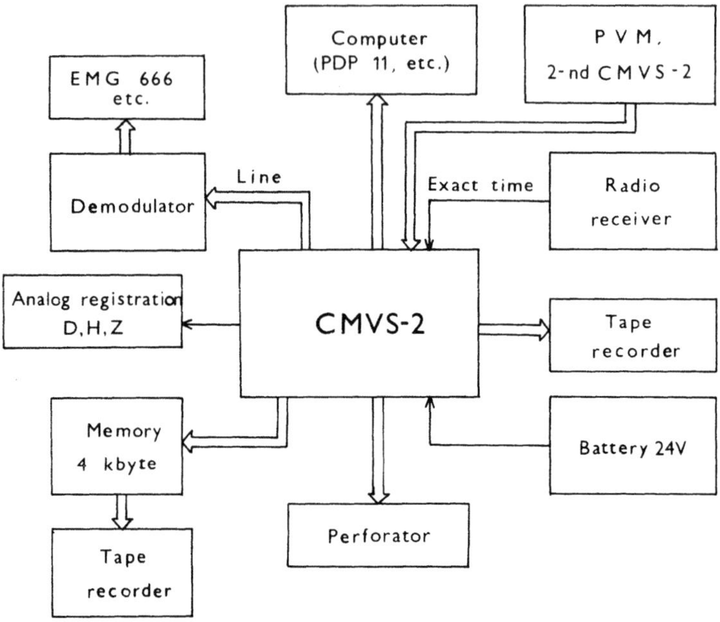

Fig. 1. Block schematic diagram illustrating the possible connections between the CMVS 2 and various external devices. The options implemented at the Hurbanovo are marked by asterisk.

controlled between 19.2 and 21.1 °C. Temperatures in the building which houses the measurement unit varied between 15 and 30 °C. The relative humidity ranged between 30% and 80% during the second half of the observing year.

The mean values of the level of artificial disturbance did not exceed 1 nT. However, pulses with amplitude of several nT were observed.

4. Results Obtained

Our observations were primarily concentrated on establishing the long term stability of the variometer baselines. For this purpose the CMVS 2 readings were compared regularly, twice a day, in the morning and afternoon, with the readings derived from a normal 20 mm hr^{-1} analogue record. On days of absolute measurements, we also calculated the baseline values for the individual variometers [1].

The measured baselines are shown in Figure 2. The running-in period of the equipment is reflected during the first months. All three variometers displayed a continuous change of baseline after July 13th, 1982, which on the evidence of the absolute measurements exhibits a jump. The most probable explanation for this is the occurrence of a release of elastic tension in the photoconverter [2].

The mean quadratic deviations of the observed baseline values are given in Table I (a), (b), (c). As can be seen the scatter of the values underwent a decrease with time for the *H* and *Z* variometers and slightly increased for the *D* variometer.

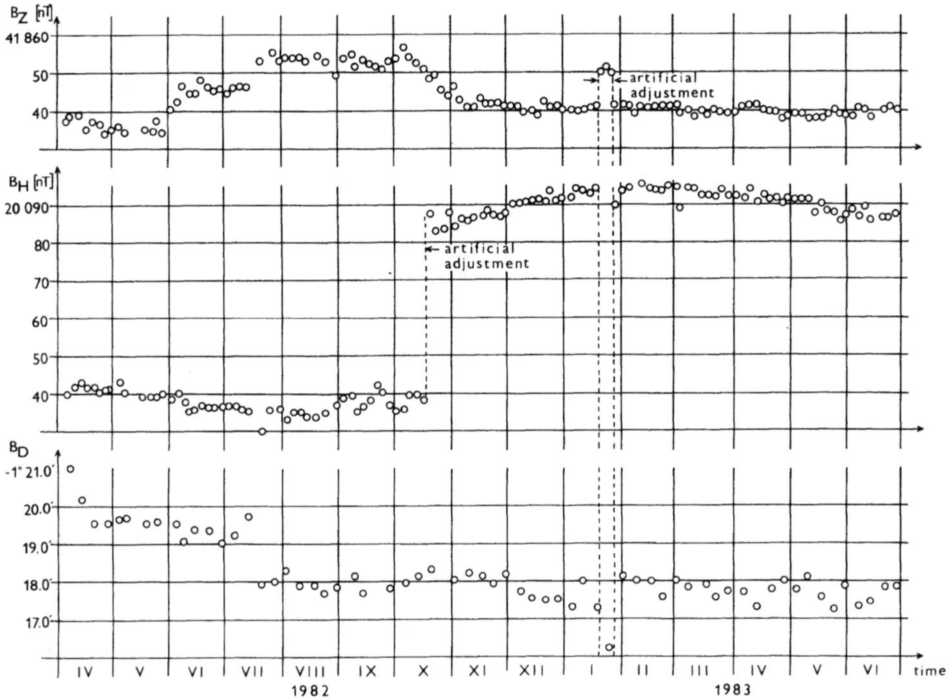

Fig. 2. Observed base-line values of the CMVS 2 magnetic variation station's *D*, *H*, and *Z* variometers.

TABLE I

Mean quadratic deviations of observed base-line values for variometers *D*, *H*, *Z*

(a)

Date	20.4–13.7.'82	14.7.–30.11.'82	1.12.'82–30.6.'83
σ_D [ ]	0.35	0.28	0.43

(b)

Date	5.4.–8.6.'82	9.6.–13.7.'82	14.7.–31.8.'82	1.9.–15.10.'82	16.10.–30.11.'82
σ_H[nT]	2.04	1.34	2.46	2.80	2.21

Date	1.12.–31.12.'82	1.1.–29.3.'83	30.3.–22.4.'83	23.4.–17.5.'83	18.5.–30.6.'83
σ_H[nT]	1.18	2.13	1.40	1.82	1.46

(c)

Date	5.4.–28.5.'82	8.6.–13.7.'82	14.7.–15.10.'82	16.10.–1.11.'82	2.11.'82–30.6.'83
σ_Z[nT]	1.91	2.07	2.07	2.10	1.37

TABLE II

Results of calibration CMVS 2 by 1000 nT (500 nT)

Date	Pulse [nT]	Amplitude [nT]			Error [%]		
		D-var	*H*-var	*Z*-var	*D*	*H*	*Z*
29. 3.'82	1000.0	1000.0	1000.0	1000.0	0.00	0.00	0.00
28. 4.	1000.0	1000.0	996.0	997.0	0.00	0.40	0.30
31. 5.	1000.0	1000.5	1000.1	999.6	0.05	0.01	0.04
24. 6.	1000.0	999.1	996.3	997.2	0.09	0.37	0.28
30. 7.	1000.0	997.1	994.8	993.3	0.29	0.52	0.67
31. 8.	1000.0	996.5	995.0	994.4	0.35	0.50	0.56
24. 9.	1000.0	998.0	993.7	993.0	0.20	0.63	0.70
28.10.	1000.0	994.8	993.2	994.3	0.52	0.68	0.57
1.12.	1000.0	997.7	996.6	994.6	0.23	0.34	0.54
29.12.	1000.0	999.0	995.8	994.6	0.10	0.42	0.54
27. 1.'83	1000.0	996.0	991.4	992.7	0.40	0.86	0.73
28. 2.	1000.0	997.5	992.8	995.4	0.25	0.72	0.46
28. 2.	500.0	499.5	497.6	498.4	0.10	0.48	0.32
8. 4.	500.0	499.0	496.7	498.3	0.20	0.66	0.34
11. 5.	500.0	499.1	497.7	497.7	0.18	0.46	0.46
30. 5.	500.0	499.7	498.1	497.5	0.06	0.38	0.50
29. 6.	500.0	499.6	498.1	497.7	0.08	0.38	0.46

The scale values were checked once a month by means of a digital multimeter with an accuracy class of 0.5. Table II shows the results obtained. It can be seen that the measured scale value, derived from the observed response to a 1000 nT (500 nT) pulse shows the least variation, 0.52%, in the case of the *D* variometer. The maximum observed percentage error for the *H* and *Z* variometers were 0.86% and 0.73% respectively.

At the present time the minute average values are recorded on C90 casettes by means of a commercially available cassette recorder. One cassette covers 34 days of recording. The recorded data can be loaded via an interface into an EMG 666 microcomputer which has a storage capacity of 8 K bytes.

Programmes are currently being developed for editing the data, and for computing and printing the hourly and monthly average values.

5. Conclusions

The results obtained indicate that the CMVS 2 is capable of meeting the stringent demands imposed on observatory instruments.

References

[1] Cifrovaya magnitovariacionnaja stancija CMVS 2, Technitzeskoye opisaniye i instrukciya po ekspluataciyi, Moscow, 1980.

[2] V. Odintsov, 1983, (private communication).

AUTOMATIC DIGITAL RECORDING OF GEOMAGNETIC ELEMENTS BY MEANS OF A PROTON PRECESSION MAGNETOMETER

A. MELONI, F. MOLINA, P. PALANGIO*, Q. TACCETTI, and ANNA DE SANTIS

Istituto Nazionale Di Geofisica, Osservatorio Geofisico Monte, Porzio Catone, 00040 Roma, Italy

Abstract. A simple system of automatic recording of geomagnetic field components by means of a proton vector magnetometer has been built and tested at L'Aquila Geomagnetic Observatory. The instrument is working with the use of a combination of the addition and compensation methods to directly measure absolute values of field components. A sophisticated equipment for the current generation and control is necessary to maintain the current flowing in the Helmholtz coils within 2 μA for successive measurements. To maximize the signal coming from the sensor inside the coils for the different components, a simple arrangement of two orthogonal small coils in only one sensor has been made.

After two years of experience and improvements the system has been further controlled comparing it with Ruska variometers regularly working in L'Aquila Observatory.

Some practical problems found in operating the system are discussed. The automatic hourly mean computation is suggested to produce yearbooks.

1. Introduction

Many Geomagnetic Observatories are still relying on optical deflecting magnets variographs to record geomagnetic components. The long process of converting the analogic traces on the photographic magnetograms in usable digitized data has been for years a tedious work in Geomagnetic Observatories.

A particular effort has been then devoted in the last 20 yr, or so, to build a completely automatic system to directly produce digital data as output (Searson, 1977).

The complete set of instruments in an Observatory includes absolute measurements instruments that are regularly used to calibrate and standardize the data coming out from the variographs. A completely automatic system which avoids absolute measurements too is then the ideal way to the optimization of a Geomagnetic Observatory. In this way a real Automatic Observatory would be made possible. This was the aim of the ASMO of Alldredge *et al.* (1964), of Yanagihara *et al.* (1973), and Andersen (1974).

Many true absolute instruments are today available to measure the total intensity of the geomagnetic field. The most used are the proton precession magnetometer (Packard and Varian, 1954) and the optical pumping magnetometer (Bender, 1960). More generally these instruments measure the intensity of any external field, thus it is possible to measure also the components of the geomagnetic field by conveniently varying the spatial configuration of the field around the sensor. Two methods can be used for this, both using an Helmholtz coils system: the compensation method (Hurvitz and Nelson, 1960), which neutralize either the vertical or the horizontal component in

* Osservatorio Geofisico Castello, 67100 L'Aquila, Italy.

Geophysical Surveys **6** (1984) 339-350. 0046-5763/84/0064-0339$01.80.

order to measure the one left; the addition method (De Vuyst and Hus, 1966), which adds external (not necessary known) fields in suitable directions.

In this paper we discuss the use of a commercial proton magnetometer to produce digital recordings of the geomagnetic elements H, Z, and D, plus total field. A preliminary report can be found in De Santis *et al.* (1981).

2. Automatic Time Sampling

A complete geometric picture of the H, Z, and D measurements is given in Figure 1. It's easy to show that adding and subtracting an extra vertical field (B in Figure 1a) and using little algebra Z can be determinated. $F0$ is the total field at time 0, when no extra field is applied, $F1$ and $F2$ are the $F1 = |\overline{F0} + \overline{B}|$ and $F2 = |\overline{F0} - \overline{B}|$. Then it follows:

$$Z = \frac{F1^2 - F2^2}{\sqrt{8(F1^2 + F2^2 - 2F0^2)}}\,. \tag{1}$$

H can immediately follow from $H = \sqrt{F0^2 - Z^2}$. To measure Declination D an extra horizontal field must be used: B' (in Figure 1b) is added once positive (towards East) once negative (towards West). It follows immediately that being $F3 = |\overline{F0} + \overline{B'}|$ and $F4 = |\overline{F0} - \overline{B'}|$,

$$D = \arcsin \frac{F3^2 - F4^2}{\sqrt{8H^2(F3^2 + F4^2 - 2F0^2)}}\,. \tag{2}$$

The time sampling for each $F(i)$ reading with a standard proton magnetometer is $\sim$4 sec; total time sampling ($F0 \ldots F4$) to obtain the components is $\sim$16 sec. The extra fields B and B' are generated by a system of vertical and horizontal Helmholtz coils, whose diameter is about 0.8 m.

Formulas 1 and 2 are exact only supposing that: (a) the extra fields (B and B') are exactly equal when added and subtracted (even if not known), (b) the levelling of vertical coils is well done, (c) for D measurements the axis of the horizontal coils is exactly in the East-West direction and well levelled, (d) the geomagnetic field does not change appreciably during the 5 field samplings. We will examine the four points.

(a) The constancy of the coil current generating the extra fields is a crucial point. It has been necessary to ensure a fluctuation not larger than 2 μA between successive samplings. To this aim a stabilized power supply has been built (Figure 2). The control stability of currents is obtained comparing the voltage at Rp with a standard voltage.

(b) The prelevelling of the vertical system is done mechanically. Magnetic levelling (Hurvitz and Nelson, 1960) gives an uncertainty of $\pm 0.1'$, then errors on H and Z are ± 1.2 nT. Horizontal coils are levelled and oriented in the East-West direction: in the realistic case of unlevelling not larger than $0.2'$ errors on D are negligible. A constantly precise levelling of the coils system is probably the principal difficulty to the use of the system as a true Automatic Observatory. It's practically impossible not having small ground movements and then coils base movements; even if small movements were not

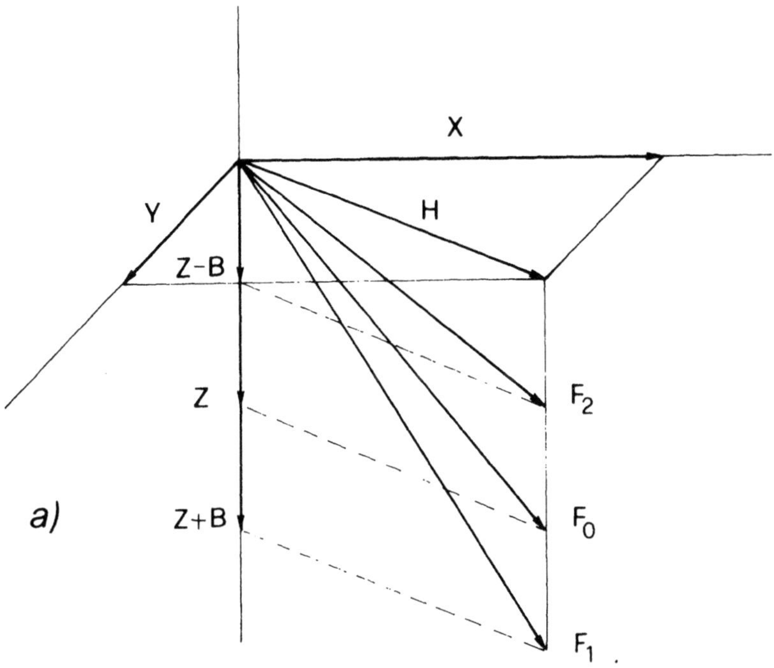

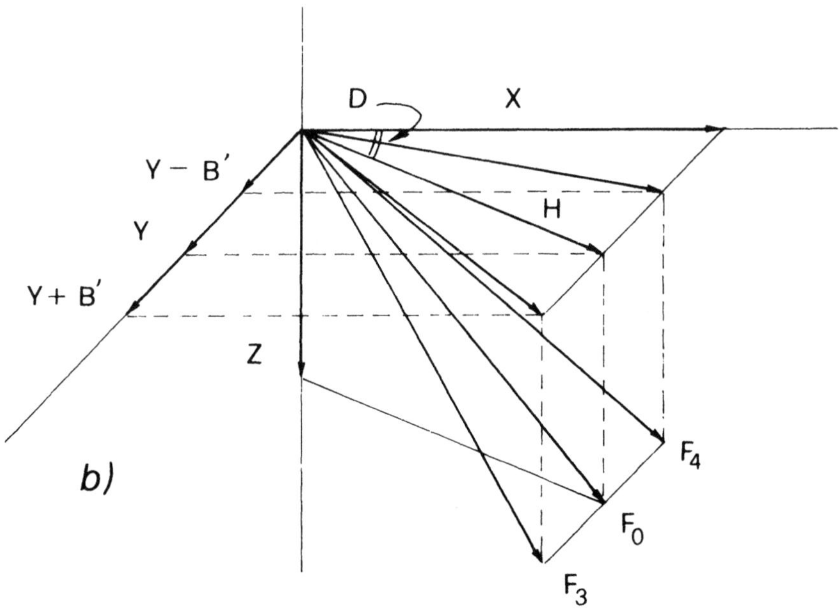

Fig. 1. Geometric picture of F, H, Z, and D measurements with additional fields: B for H and Z measurements, B' for D measurements.

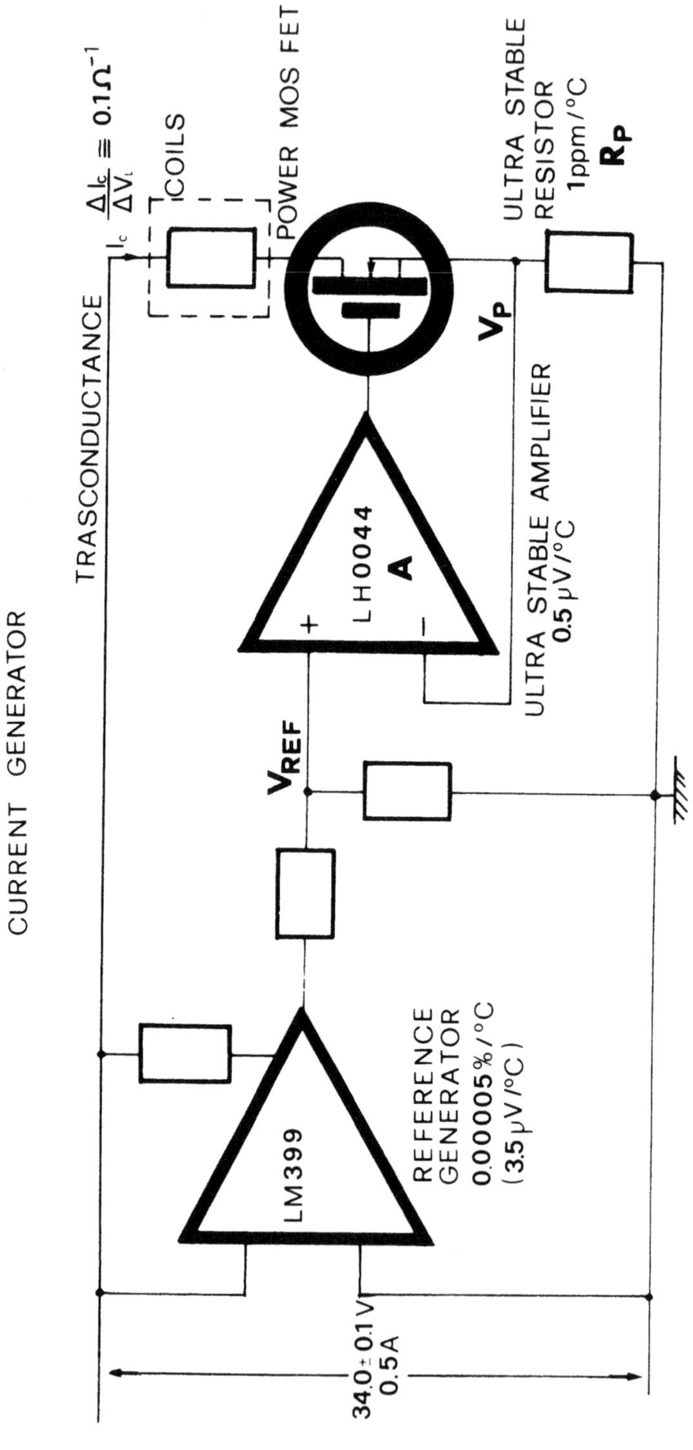

Fig. 2. Stabilized Power Supply electronic schematic.

detected the importance of checking the stability of the system through comparison with independently taken absolute measurements should not be minimized. This check was done for L'Aquila's Observatory (see Discussion).

(c) Being very difficult to precise orient the horizontal coils axis of the system in the geographical East–West direction, in (2) the angle D must be replaced by an angle $D' = D + \alpha$, where α is the exorientation angle with respect to the East–West line; this quantity can be computed only once through independent simultaneous absolute measurements of Declination. It's however important to check this 'instrumental constant' periodically.

(d) A total time sampling of 16 sec is a very small part of the trace in normal running variographs ($\sim$0.1 mm on variographs with a time scale of 20 mm hr^{-1}); therefore if measurements are made regularly they are generally sufficient to plot out the right time variation of the geomagnetic components. Some geomagnetic variations are however very rapid; excluding micropulsations, we should take in account the SSC's and SI's: these sharp movements, generally well detected in photographic variographs, can rise 0.3 nT sec^{-1} at medium latitudes and more at high latitudes; the no contemporaneity of the 5 measures can in this case be a major problem. It can be noted that a reduction of the total sampling time is possible at medium latitudes. If the **B** upward positive field (Figure 1a) is made equal to the vertical component, $F2$ will coincide practically with H. The current needed to compensate Z is not critical at all; for example a compen-

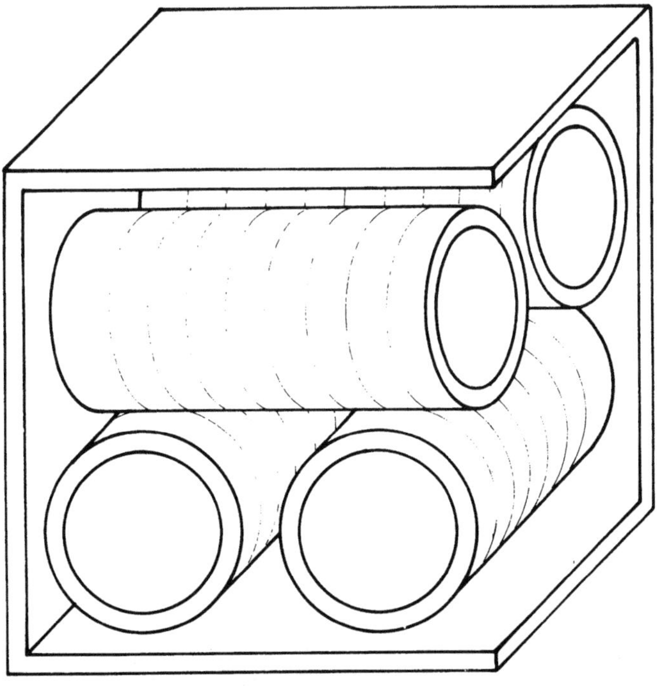

Fig. 3. Drawing of Automatic System sensor with two groups of two coils mutually perpendicular.

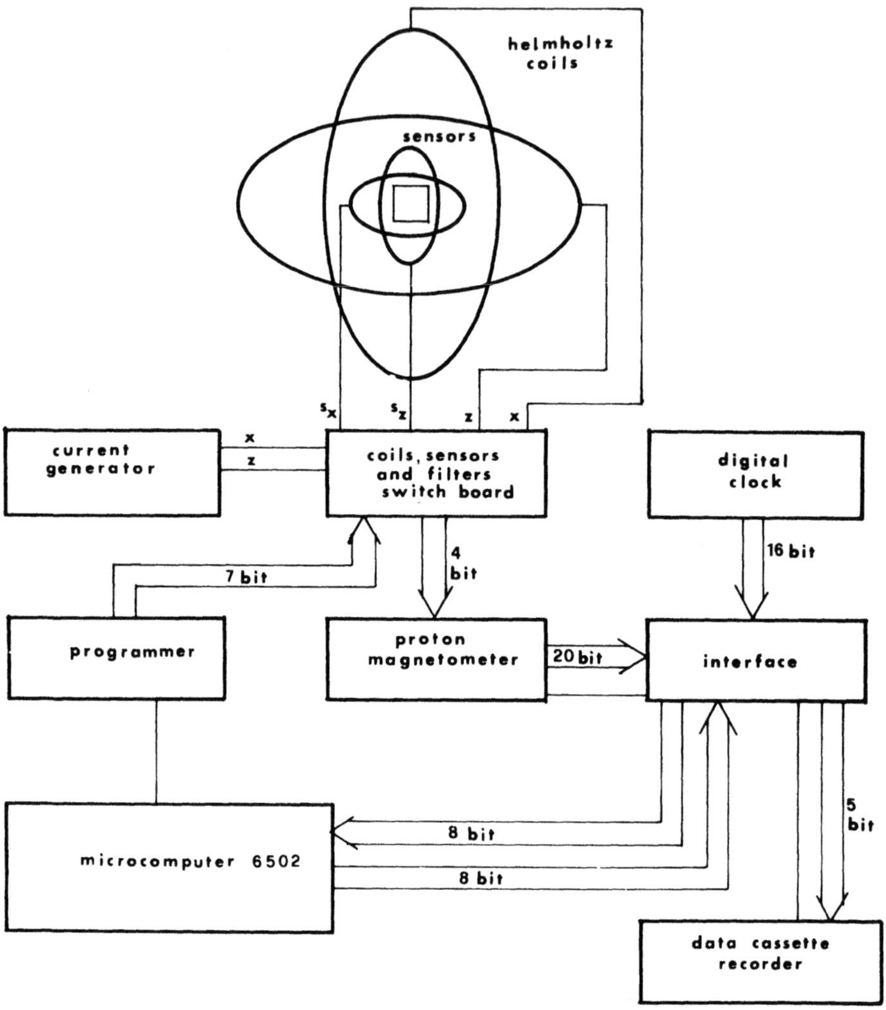

Fig. 4. Block diagram of Automatic System.

sation error on Z of ~ 100 nT gives at L'Aquila an error of 0.2 nT on H. This peculiarity can be used to further reduce the total time sampling; the sequence can be $F0$, $F2 = H$, $F3$, $F4$, $(F1)$. $F1$ can now be used only as a control. Using this last sequence the error on H is practically the same as on $F0$ (± 1 nT), Z is then affected by an error of ± 1.3 nT and D of $\pm 0.16'$. $F0$ and H are instantaneous values, Z can be considered a mean value in the 4 seconds between $F0$ and H samplings; D is practically a mean of $F3$ and $F4$.

The signal to noise ratio in a proton magnetometer is proportional to the square sinus of the angle between the direction of the external field and the direction of the sensor coil axis. To maximize the S/N ratio the sensor coil should always be put at right angle with the external field. In our system the angle between the optimum directions of the sensor coil axis for Z (or H) measurement and D measurement is near 90°; to avoid any mechanic device to rotate the coil axis a particular sensor was designed. In Figure 3 a drawing of the Automatic System sensor is shown; four solenoids are soaked in an hydrogenatum liquid, forming two groups of relatively orthogonal coils. In order to have a large number of coils in a small space, the two groups are made of two solenoids each; the two groups are respectively switched according to the field component that needs to be measured. In the actual case the total volume occupied by the box containing sensors and liquid should be large enough to obtain an acceptable S/N ratio and small enough to avoid magnetic gradients in the measuring volume. In our case a $15 \times 15 \times 18\,\text{cm}^3$ box has been used.

The precession magnetometer is not affected by temperature variations but the system of compensating Helmholtz coils can slightly change its geometry with temperature gradients between the different parts of the system. To avoid any temperature effect the coils system, and sensor inside, have been closed in a (approximately 1 m) styrofoam box 0.12 m thick.

The chosen data acquisition system is a magnetic cassette recorder. A complete set of measurements at 1 min sampling rate is made of $F0$ (12 bits), H (12 bits), D (12 bits), hr and min (16 bits). A preliminary elaboration of data has been used to compact, in the above mentioned way, the output data. A microprocessor takes care of the automatic sampling and preliminary data elaboration. A C90 cassette lasts for 23 days. A complete block diagram of all system is shown in Figure 4; all instrumentations are powered at 12 V with 4.5 A. In case of power fail a normal car battery (60 A hr) can provide about 12 hrs of autonomy.

3. Discussion

After all data are recorded on the digital device it's easy with any computer to develop a software to handle the data in an effective way.

The very first result is an analog reconstruction of the normal daily magnetogram; this can be done with a plotter device. In Figures 5 and 6 two days in 1982 are shown with the plotter reconstruction in the lower part of the figure and the Ruska magnetogram in the upper part. All Automatic System 1 min samplings are reported with no filtering process applied on data. In Figure 5 the plotter vertical axis scale has been adjusted to have the same Ruska scale value for that day, while in Figure 6 the plotter scale has been enhanced; these figures show the capability of the System to follow rapid geomagnetic variations too. This result is obvious for the H component since it has been measured instantanously. Declination comes from a computation over the time interval of 12 sec but as can be seen this does not affect the reconstructed trace. It should be necessary to test however single rapid variations (like SSC or SI) but we think that this check could be affected more by the 1 min sampling process (adopted in this initial configuration of our Automatic System) than by the non simultaneity of the measurements.

Figures 7 and 8 show results of a long term stability test of the System in the March–

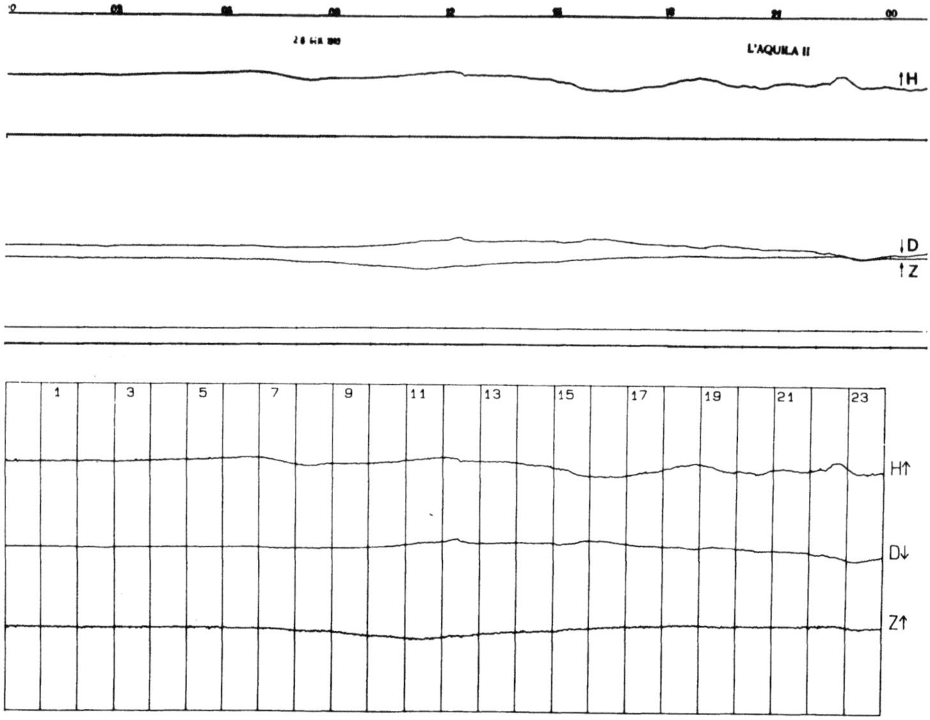

Fig. 5. January 28th, 1983: upper part L'Aquila Ruska magnetogram and lower part plotter reconstruction from Automatic System data.

December 1982 time interval. Differences in Z, H, and F, computed averaging in the time interval when absolute measurements were taken, are shown in Figure 7. In Figure 8 02 U.T. instantaneous differences are reported for Z, D, and H; in this case the Automatic System values are compared with magnetograms for which base line and scale values were chosen after smoothing and interpolation between absolute measurements was done. The high average difference as reported on the y-axes on Figures 7 and 8 is due to the fact that the Automatic System is physically located in a different building; this difference is only the ground anomaly of the two locations.

The figures 7 and 8 are essentially an indication of the technique used at L'Aquila Geomagnetic Observatory for the data elaboration: (1) by means of absolute measurements the Automatic System stability is checked or, in other words, its base lines are computed; (2) once this check is done, the Ruska variograph drifts are monitored day by day by the Automatic System also for days between absolute measurements. In Figure 7, which corresponds to point 1, a little slow and regular drift in H and Z from March to September (days 30–270) is clearly recognizable; this drift amounts to ~ -5 nT in H and $\sim +3$ nT in Z. The probable reason for this drift is a slow unlevelling of the Helmholtz coils due to ground movements. During the last part of the year base lines of the Automatic System vary more rapidly and not regularly; this was caused in somes cases by errors in absolute measurements. The Automatic System has then contributed to judge possible mistakes in single absolute measurements. It's however sufficient to give a glance to the Ruska variograph base line during the same months of Figure 7 (not shown here) to conclude that the Automatic System is decidedly more stable.

Figure 8 shows, for Z and H, a general picture almost equal to the one shown in Figure 7; this was expected since base lines definitely adopted for Ruska variographs are deduced from the same absolute measurements from which Automatic System base lines are computed. Superimposed to a general trend spikes and fluctuations, whose cause can be found only in some cases, can be seen. For example at day 150 the sudden jump in H and Z is due to a thermic drift of the Ruska variographs (twice a year thermostatic control level of recording room is changed). Other fluctuations like the oscillation of H in the first half of July (after day 181) could be justified with slight movements of the coils system but this should cause similar fluctuations in Z that are not seen. It's probably more realistic to think that the Ruska H variograph base line has fluctuated (the one adopted is constant from July 2nd through the 21st). Similar problems can cause fluctuations on D. However small fluctuations like those seen in these plots can be easily smoothed for a yearbook compilation and new base lines can be adopted.

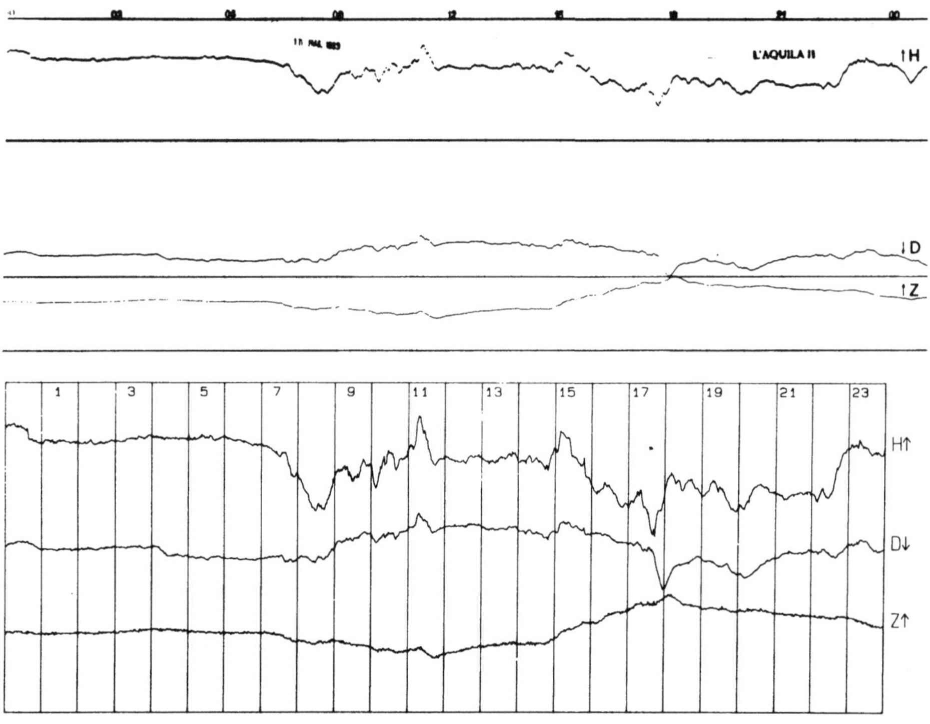

Fig. 6. May 11th, 1983: upper part L'Aquila Ruska magnetogram and lower part plotter reconstruction from Automatic System data with enhanced scale constant.

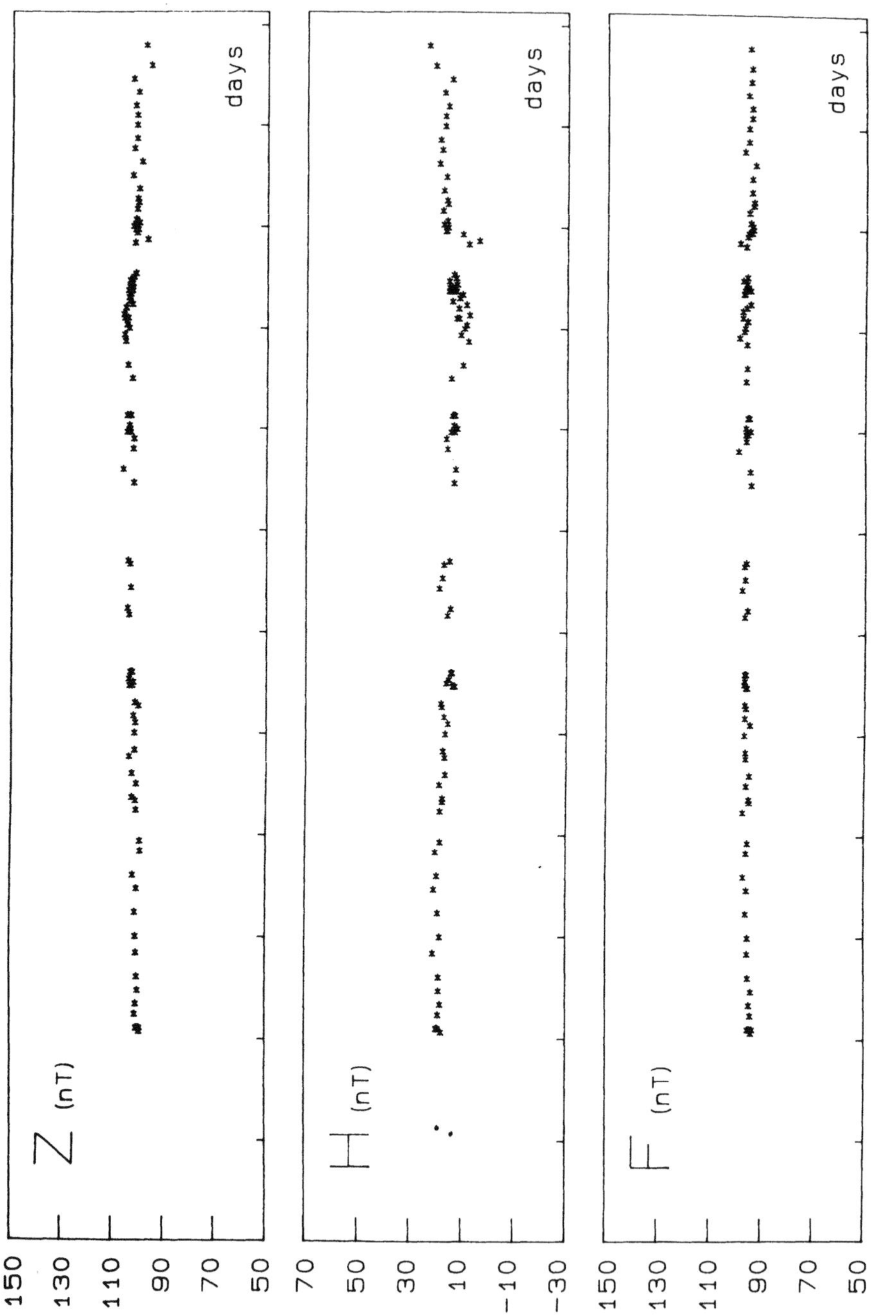

Fig. 7. Automatic System values minus Absolute measurements for *Z*, *D*, and *H* in the March–December 1982 time interval.

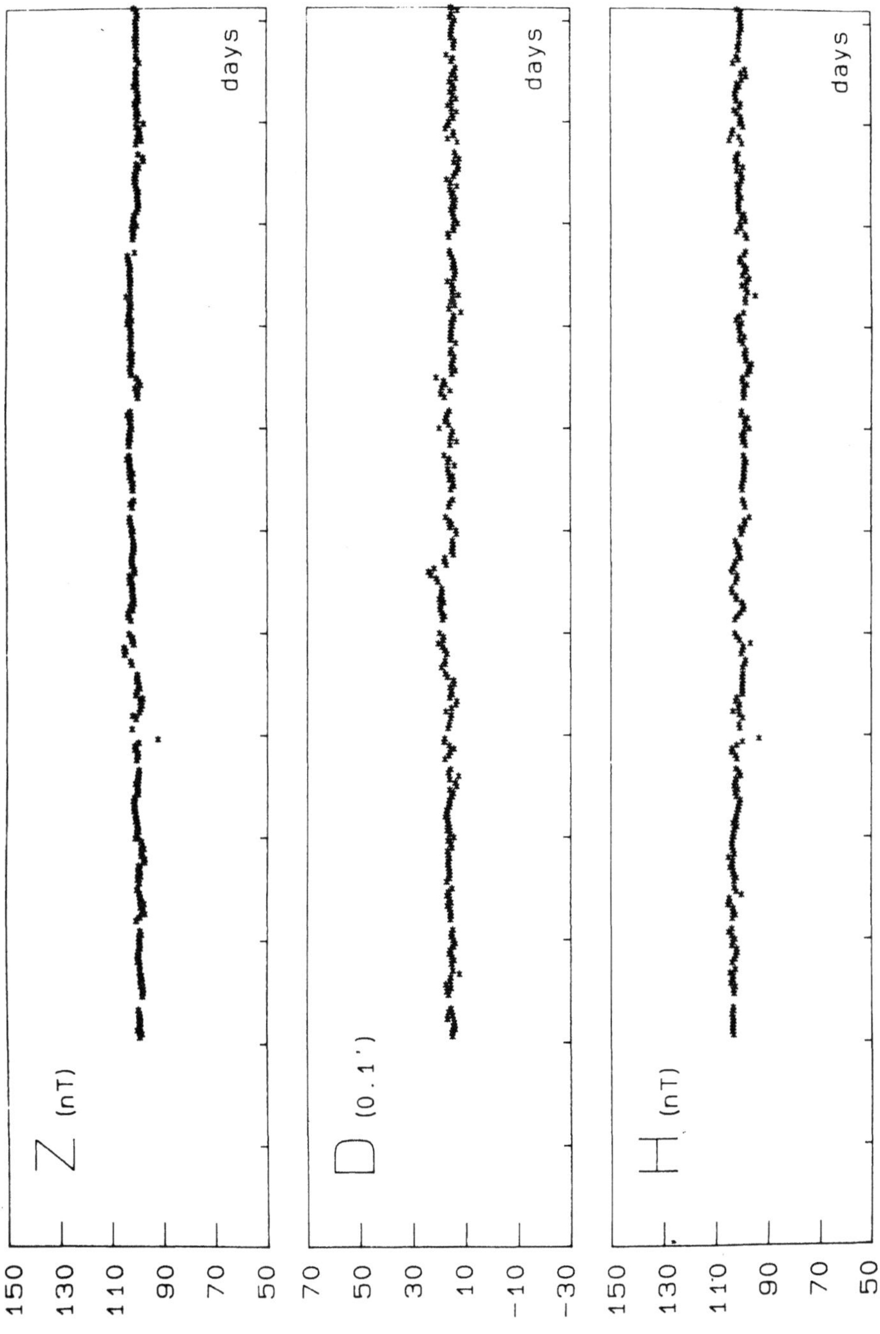

Fig. 8. 02 U.T. instantaneous differences of Automatic System values minus Ruska values for *Z*, *D*, and *H* in the March–December 1982 time interval.

4. Conclusions

(1) After two years of experience we can state that the Automatic System is much more stable than normal photographic paper variographs. A disadvantage is certainly due to the possibility of having interruptions and fails in the regular operation of the System. The real danger for the continuous recording is due to lightnings in the area of the Observatory.

(2) The digital output can be used for hourly means computations to produce yearbooks automatically; this has not yet been done at L'Aquila systematically but it has been used to fill possible holes in photographic recordings. The situation will eventually be inverted.

(3) The Automatic System is not a real Automatic Observatory: as it has been shown in the previous sections, absolute independent measurements are most valuable even if Automatic System is much more stable than normal variographs. To reach an ideal situation ground movements should be tested and possibly recorded in order to correct, when it's the case, the components measurements accordingly (clinographic recorders will early be used). If however drifts are also due to other causes like slight unpredictable deformations of the Helmholtz coils the possibility of making any correction seems unrealistic.

Acknowledgments

We thank L. Innocenzi for the photographic work.

References

Alldredge, L. R. and Saldukas, I.: 1964, 'An Automatic Standard Magnetic Observatory', *J. Geophys. Res.* **69**, 1963.

Andersen, F.: 1974, 'An Automatic Magnetic Observatory System', *Publ. of Earth Physics Branch, Ottawa, Canada* **44**, 11.

Bender, P. L.: 1960, 'Measurements of Weak Magnetic Fields by Optical Pumping Methods', *Bull. Ampere* **9**, 621.

De Santis, A., Meloni, A., Palangio, P., and Taccetti, Q.: 1981, 'Un sistema automatico per la misura di tre componenti del campo Geomagnetico utilizzante un magnetometro a precessione nucleare', *Annali di Geofisica* XXXIV, 83.

De Vuyst, A. and Hus, J. J.: 1966, 'Generalization de la mesure de l'intensite des composantes verticale et horizontale du champ magnetique terrestre avec le magnetometre a protons', *Annal. de Geophys.* **22**, 119.

Hurvitz, L. and Nelson, H. J.: 1960, 'Proton Vector Magnetometer', *J. Geophys. Res.* **65**, 1759.

Packard, M. and Varian, M.: 1954, 'Free Nuclear induction in the Earth's Magnetic Field', *Phys. Rev.* **93**, 941.

Searson, P. H.: 1977, 'Digital Recording and Telemetry of Magnetic Observatory Data', Talk given at IAGA, Seattle U.S.A.

Yanagihara, K., Kawamura, M., Sano, Y., and Kuboki, T.: 1973, 'New Standard Magnetic Observation System of Kakioka', *Geophys. Mag.* **36**, 217.

COMPARISON OF STANDARD AND DIGITAL OBSERVATION AT THE GEOMAGNETIC ADOLF SCHMIDT OBSERVATORY NIEMEGK

V. AUSTER and K. LENGNING

Central Institute of Solar-Terrestrial Physics, Acad. Sci. G.D.R., G.D.R.-1199, Berlin

Abstract. For more than 4 yr the geomagnetic field has continuously been recorded by means of 3 proton magnetometers (F, X, Y) at Niemegk Observatory (1 value/min on punched tape). With this observation material, for all 7 geomagnetic components hourly means are calculated. The preliminary hourly means for F, Y, and X are published in the monthly report. As a rule, their comparison with the observatory data is satisfactory. Only with great geomagnetic activity ($K_1 \geq 7$) what was the case especially in 1982 relatively often, deviations in the hourly means up to 30 nT occurred. Obviously, during great magnetic disturbances the scanning rate of 1 min is not sufficient. In case of quick data variations, however, small technical defects in the electronic circuit can occur. Besides, these great differences could also be explained by errors which could be conditioned by the standard observational and processing methodology.

Since 1972, at Niemegk Observatory tests with regard to an establishment of a digital geomagnetic recording with nuclear resonance magnetometers have been carried out (Auster *etal.*, 1972) and (Auster, 1974). For these recording systems, as variometers one proton magnetometer is used for recording F and two component proton magnetometers are used for recording X and Y. The minute values for F, X, and Y are recorded on punched tape. From these 60 values hr^{-1} in each case the hourly means are calculated by computer, and from them the other 4 geomagnetic components D, H, Z, and I, so that all 7 geomagnetic components can be compared with the geomagnetic observatory data what ensures a further improvement of accuracy of digital recording.

In the Monthly Report of Niemegk Observatory the results have been published since January 1979. For technical reasons of processing, the monthly schedules for F, X, and Y arranged in the ordinary manner are published only in the following month. One part of these schedules is shown in Table I. Above the tables with the preliminary hourly means (F, X, Y) of the digital recording there are the basic value and the variation around this value in the corresponding month. These values ΔF, ΔX, and ΔY give a good survey regarding the zero point constancy of the proton variometers. ΔF, ΔX and ΔY for the 5 yr from 1979 till 1983 are contained in Figure 1.

For each month, only on 4 selected days the digital recording is compared with the recording results of the observatory (magnetic standard recording with 3 systems).

For this purpose, instantaneous values and hourly means are used. From this comparison the zero point variations ΔF, ΔX, and ΔZ are derived. Naturally, first of all the pilar differences are contained in them. Since the installation differences were rather great until the end of 1982 (digital recording to the absolute house nearly 100 m), ΔF values of ± 3 nT are resulting. The rather great ΔX and especially ΔY values can be

Geophysical Surveys **6** (1984) 351-355. 0046-5763/84/0064-0351$00.75.

TABLE I

Magnetic total intensity Niemegk $F = 48\,000\,\text{nT} + 1\,\text{nT}$ + tabular values in nT
Preliminary hourly mean values April 1982

U.T.00	01	02	03	04	05	06	07	08	09	10	11	12
1	419	419	418	418	420	424	427	424	415	406	398	397
2	396	379	379	383	391	399	408	405	400	395	390	386
3	407	406	402	405	402	400	404	415	415	399	390	394
4	419	409	408	409	414	415	420	422	419	408	407	405
5	420	414	417	416	414	419	422	417	402	393	385	393
6	420	424	421	422	425	430	430	419	412	396	388	390
7	419	419	420	422	424	426	427	422	408	392	384	387
8	427	426	427	427	428	427	424	415	406	395	391	396
9	420	419	420	418	420	423	423	418	406	395	387	388
10	427	423	421	420	420	419	419	415	403	386	378	383
11	409	401	399	399	406	409	401	399	400	388	381	380
12	429	429	429	429	430	434	433	427	415	402	390	391
13	428	427	424	413	416	425	424	418	405	394	386	386
14	427	427	428	428	430	433	433	428	417	401	389	388
15	428	427	427	427	428	430	430	425	412	395	386	382
16	424	425	425	424	429	430	429	424	410	394	387	386
17	431	432	427	424	429	433	432	421	404	392	393	402
18	426	425	426	427	428	428	424	414	404	391	383	382
19	424	425	426	427	428	429	427	421	410	399	390	388
20	421	420	419	424	421	424	426	421	406	403	401	403
21	422	412	418	425	428	430	426	415	409	404	396	387
22	408	404	403	409	406	414	420	421	414	407	393	386
23	425	425	425	426	426	426	424	416	409	404	398	397
24	424	424	423	424	428	425	417	414	409	402	395	393
25	435	429	414	415	405	408	413	409	383	371	376	385
26	426	425	425	425	425	426	422	411	402	396	388	390
27	421	424	425	425	426	423	415	411	408	398	392	395
28	419	415	404	408	408	410	413	411	411	404	386	385
29	417	418	419	422	421	424	418	411	402	397	389	389
30	409	400	407	413	411	411	410	408	410	396	384	388

explained by the very complicated component magnetometers (electronic equipment and adjustment). Failures of digital recording have been replaced by values of observatory recording. In most cases, these failures are caused by the electric mains (mains failure, switchings), but also by near thunderstorms and errors of the equipment. ΔF, ΔX, ΔY are taken from the Monthly Reports. As to be expected, the variations from month to month are lowest in ΔF. The most significant variations from July 1981 (Nov. $+4$ nT) are caused by works at the devices. In the end of 1982, a new recording system was installed (little disturbance at the end of series ΔF, ΔX, and ΔY). This 2nd digital recording system is now the main system. It is installed in a small observation pavilion situated northwest from the absolute measurement pavilion. The annual course as it could be observed between the proton magnetometer measuring

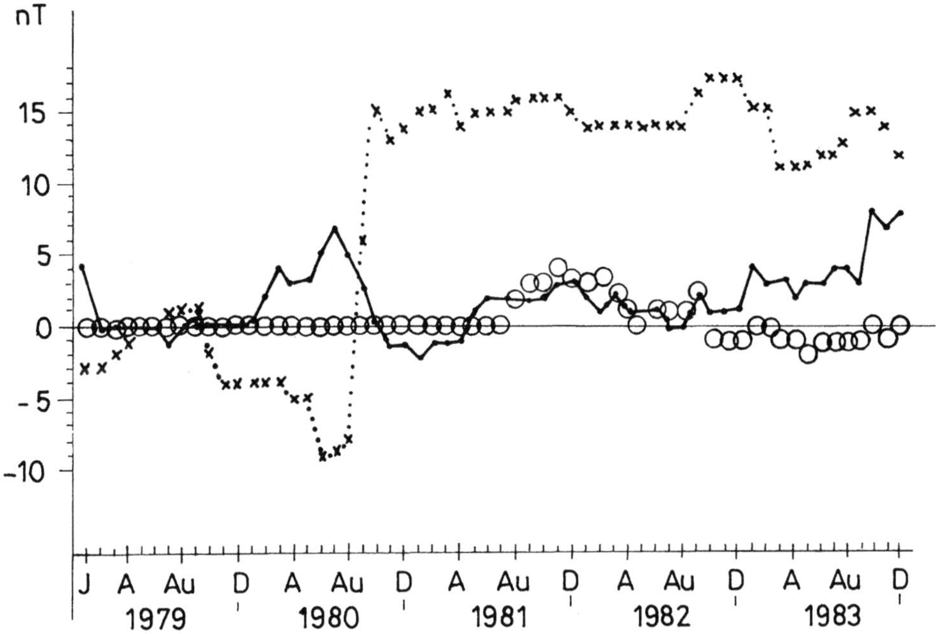

FIG 1

places 'Absoluthaus' and 'Waldpfeiler' already for years (Lengning *et al.*, 1968), does not occur any longer.

This foundation difference in F is not constant and has a seasonal variation of approximately 2 nT. The variation measuring pavilion is electrically heated and the 'Waldpfeiler' measuring foundation is situated near the power supply cable.

ΔX, too, shows relatively small deviations from the basic value what is unfortunately not valid for ΔY. In the period of 5 yr discussed here the ΔY covers a range of -9 to $+17$ nT. After a favourable beginning, greater deviations to negative values occurred so that in August 1980 an intervention was made which caused an unintentional great rise. In September the value for ΔY is about $+6$ nT and reaches in October $+15$ nT, though no further readjustment was carried out. With unsignificant variations, this value has then been kept on a level near 15 nT. The continuous comparison with classical standard measuring must still be continued in the next years.

In Table II, for reasons of saving place only the annual means of F, X, and Y and that month of the year which has the greatest deviations have been compiled. Additionally to the differences to the observatory mean values, the annual and monthly A_K are contained in the table, too.

For this comparison, on the one hand the values of preliminary hourly means tables for F, X, and Y from the Monthly Reports are used in which the variometer variations are naturally not contained any longer. F, X, and Y have been matched to the preliminary basic data of observatory results. On the other hand, for this comparison in Table II the final observatory results are used. In Table II the results of both different

TABLE II

Year month	Annual mean dig. rec. monthly mean dig. rec.			Difference to the observatory means			Magn. activity yearly A_K monthly A_K
	F_D nT	X_D nT	Y_D nT	$F_D - F_{Obs.}$ nT	$X_D - X_{Obs.}$ nT	$Y_D - Y_{Obs.}$ nT	
1979	48364.8	18816.2	−228.0	+1.7	−0.3	+0.2	12.9
January	48357	18805	−243	+3.4	−0.4	0.0	15.8
1980	48379.8	18823.1	−191.9	+2.4	+0.5	+0.1	10.1
April	48372	18826	−199	+3.5	+1.0	−0.4	9.9
1981	48404.8	18816.3	−153.8	+0.7	0.0	−0.1	16.2
July	48404	18824	−152	+2.8	−0.3	−0.5	16.8
1982	48425.7	18807.1	−115.9	0.0	−0.8	−0.2	18.7
September	48432	18800	−106	−2.4	−10.7	+0.5	26.5
1983	48440.0	18809.1	− 82.1	+0.2	−0.3	−0.2	16.9
May	48438	18816	− 87	+1.9	+0.5	+0.5	18.5

recording methods are compared. The differences can be caused not constant foundation differences and evaluation errors in standard and digital recording.

Obviously, the differences of the annual means are satisfactory. Only for F, the years 1979–81 always have positive values around 1–2 nT for which the place of instalment 'Waldpfeiler' is considered to be responsible for from 1982 this difference became significantly smaller. The proton magnetometer measurements of magnetic total intensity F for absolute determination have been carried out already for 20 yr in the absolute pavilion.

The monthly means with the most significant deviations in one year always occur with the magnetic total intensity, too. In 1982, however, the X component has the largest value. It is here a question of the month September with an A_K of 26.5. The mean K_1 of this month reached 3.35 so that in the standard method larger processing errors are possible, too.

Naturally, for such a strong variation activity a scanning rate of 1 min for hourly means formation should not be sufficient. In February 1982, the mean K_1 was even 3.46. (This month was in the Niemegk measuring series since 1890 the February with the greatest magnetic activity.)

The putting into operation of a 2nd digital recording system has already decreased the failure rate. In order to avoid the difficulties connected with the scanning rate of 60 values hr^{-1}, the 2nd recording system is indeed to be operated with a scanning rate of 60 values hr^{-1}, too, but the operating time is shifted for 30 s. Thus, during magnetic disturbances for processing both digital systems can be used and per hour 120 data are available. In these 5 yr, in 2 cases this comparing control has proved usefull for the evaluation of magnetic standard registration. The new magnetic observation method with vector proton magnetometers and with digital output is continuously being improved. Especially for the technical prosecution of recording and for the new evaluation methodology further experience is necessary. The approved standard observations are to be continued in the same manner for the necessary comparing control must be carried out in future, too. Probably in 1–2 yr it will already be possible to do without the very extensive graph reading (determination of hourly means by analogue recordings). The diurnal determination of additional instantaneous values is then increased necessary. Using the magnetic nuclear resonance theodolite (Auster, 1984) for absolute measurements this new magnetic observatory practice can be protected and improved quite significantly with regard to accuracy because then the number of yearly available absolute measurements for F, Z, X, Y increases for a factor of nearly 10. Corresponding tests have already successfully started at Niemegk Observatory.

Finally it should be stated that in this short consideration the preliminary means of digital recordings have been compared with the results of the observatory determined only after the end of the measuring year.

References

Auster, V.: 1974, 'Zum Langzeitverhalten von Kernresonanz-Kompensationsvariometern', *Jahrbuch d. Obs. Niemegk*, S. 132–134.

Auster, V.: 1984, 'Geomagnetic Absolute Measurements with a Nuclear Resonance Theodolite at the Adolf Schmidt Observatory in Niemegk', *Geophys. Surveys* **6**, 269-270 (this issue).

Auster, V., Lengning, K., and Schmidt, H.: 1972, 'Kurze Angaben zu dem Projekt der Erfassung und Auswertung digitalisierter Beobachtungsdaten am Observatorium Niemegk', *Jahrbuch*, S. 132–137.

Lengning, K., Schmidt, H., and Zander, W.: 1968, 'Ergebnisse der Beobachtungen am Adolf-Schmidt-Observatorium für Erdmagnetismus in Niemegk im Jahre 1968', *Jahrbuch d. Obs. Niemegk*, S. 25.

IMPROVED KAKIOKA AUTOMATIC STANDARD MAGNETOMETER (KASMMER)

M. KUWASHIMA and Y. SANO

Kakioka Magnetic Observatory, Japan Meteorological Agency, 595 Kakioka, Yasato machi, Niihari-gun Ibaraki, 315-01 Japan

Abstract. The KASMMER (Kakioka Automatic Standard Magnetometer) has been constructed in 1972 for the standard magnetic observation at Kakioka (Yanagihara *et al.*, 1973). Then KASMMER system has been improved successively for the high level standard magnetic observation. The purpose of the present paper is to introduce an outline of the improved KASMMER system.

The KASMMER system consists of four parts which are four optical pumping magnetometers, a fluxgate magnetometer (Supporting system to the optical pumping magnetometers), a calibration system and a data acquisition system (computer system). The calibration system consists of a proton magnetometer and a magnetic theodolite (Di–72). Details for each part are described and the ability of the improved system is shown to be higher than that of classical magnetic observatory systems.

1. Optical Pumping Magnetometer

The block diagram of the optical pumping magnetometer is shown in Figure 2. Four optical pumping magnetometers measure total force (F), horizontal component (H), vertical component (Z), and a special component (Hy) which is converted to D-component. The measurement is carried out with an interval of one second and with a resolution of 0.1 nT. The optical pumping magnetometers employ the cesium oscillator

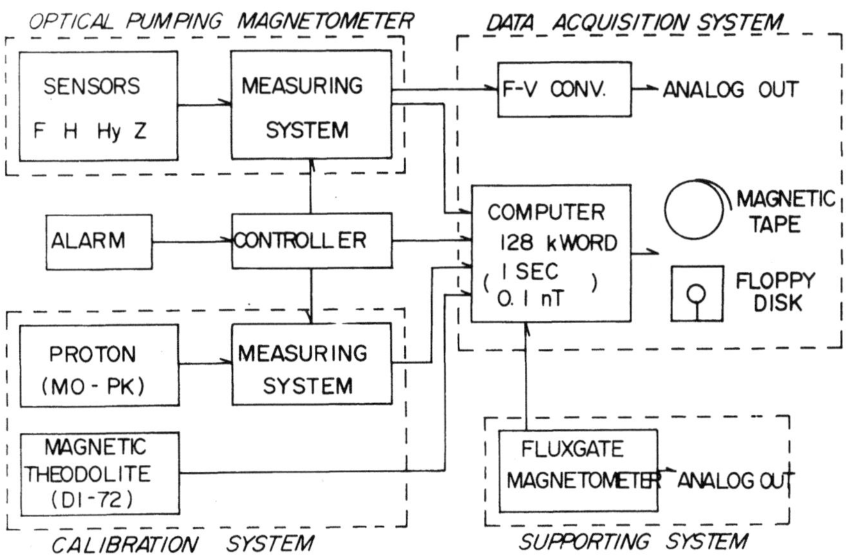

Fig. 1. Block diagram of the KASMMER system which consists of four optical pumping magnetometers (F, H, Z, Hy), calibration system, data acquisition system and fluxgate magnetometer.

Geophysical Surveys **6** (1984) 357-365. 0046-5763/84/0064-0357$01.35.

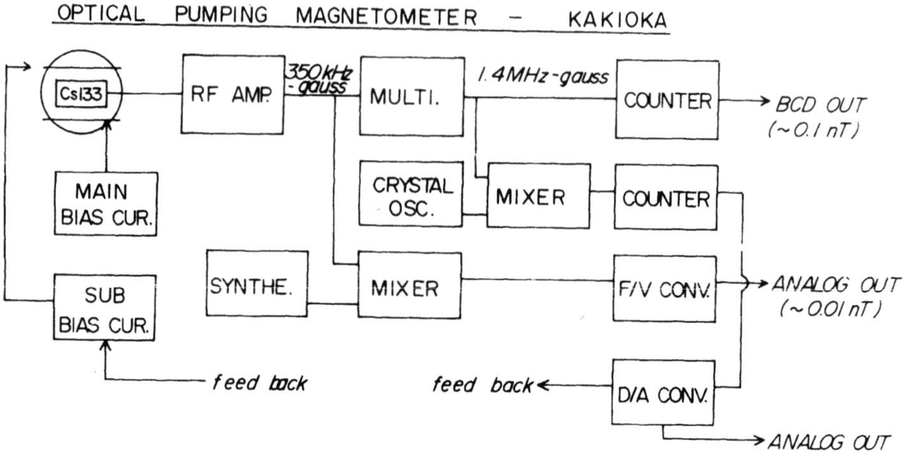

Fig. 2. Block diagram of the optical pumping magnetometer.

as the sensing element to produce an output signal whose frequency is proportional to the ambient magnetic field intensity. The cesium sensor is laid at the center of the helmholtz coil pairs which produce a suitable bias field to measure a component. The frequency from the cesium oscillator is multiplied by 4. The electronic counter counts the multiplied frequency upon 1/1.4 sec gate, which is to give directly the nT value of the measured magnetic field in 0.1 nT unit. The employed conversion constant, 1.4 MHz G^{-1} comes from the Larmor frequency of cesium, 350 kHz G^{-1}, in a weak magnetic field.

The sensing element of the optical pumping magnetometer is laid at the center of the two orthogonal helmholtz coil pairs. Each helmholtz coil pair has to be oriented in a proper direction and its coil current has to be regulated to give a proper bias field for a specified component of the geomagnetic field to be measured. In the *D*-component optical pumping magnetometer of KASMMER, its sensor detects a force components of Hy in the horizontal plane as shown in Figure 3. The direction of the Hy makes an angle of 60° eastwards from the mean magnetic meridian. The perpendicular components Hx in the horizontal plane and Z-component are eliminated by using the two orthogonal helmholtz coil pairs. A calculation of the *D*-component is carried out by the following relation,

$$D = \cos^{-1}(Hy/H) - \theta_0$$

where θ_0 is about 60°.

2. Fluxgate Magnetometer

In the improved KASMMER system, a fluxgate magnetometer is installed as a supporting system to the optical pumping magnetometers. It is well known that a fluxgate magnetometer has some problems, which are the drift effect and large

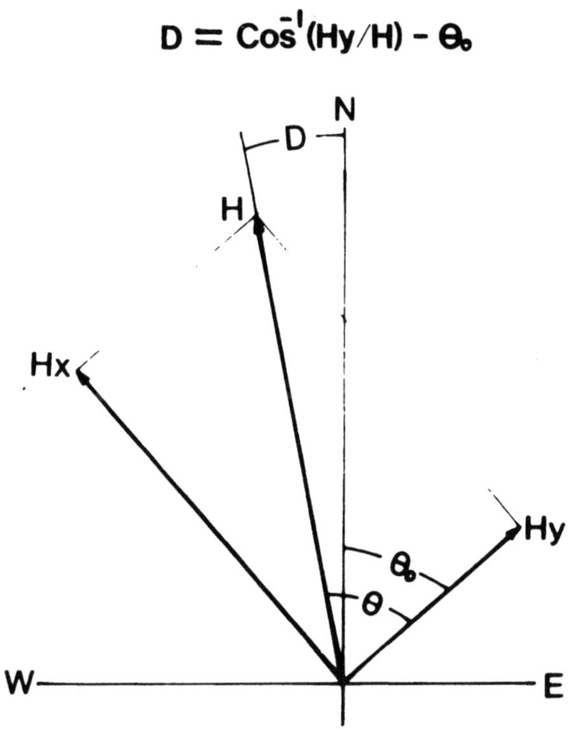

Fig. 3. Direction of *Hy* which is used for the observation of *D*.

temperature effect. However, we have improved the fluxgate magnetometer which is able to be adopted as magnetic observatory system. This newly designed fluxgate magnetometer has rejected the problems of drift and temperature effects. The improved fluxgate magnetometer can measure magnetic field with a resolution of 0.1 nT, with a very small temperature effect (0.1 nT $°C^{-1}$), and with a very small drift effect (0.1 nT day^{-1}). Therefore, the improved fluxgate magnetometer has enough abilities to support the optical pumping magnetometer.

3. Calibration System

Calibration system consists of a proton magnetometer and a magnetic theodolite (DI–72). The proton magnetometer determines the absolute values of total force (F). Generated signal is amplified and then multiplied by 200. An electronic counter counts the multiplied frequency, giving the digital value in 0.1 nT unit for 1.17437 sec determined by the gyro-magnetic ratio,

$$\gamma_p = 2.67513 \times 10^4 \, \mathrm{sec}^{-1} \, \mathrm{G}^{-1}.$$

The measurement can be repeated automatically at every 6-sec or one minute interval.

The DI–72 is a magnetic theodolite to measure declination D and inclination I, with a helmholtz coil mounted on a theodolite and a rotating search coil set at the center. The helmholtz coil creates a magnetic field $\bar{F}c$ whose intensity is approximately equal to geomagnetic field intensity $\bar{F}$. The direction of $\bar{F}c$ is also approximately equal to that of the geomagnetic field intensity. Changing the direction of $\bar{F}c$ to approach just reverse to $\bar{F}$, the transverse component $\Delta\bar{F}$ of the vector sum, $\bar{F} + \bar{F}c$, is detected by rotating search coil whose axis is lying along $\bar{F}c$. When the output AC signal of the search coil becomes zero, the direction of $\bar{F}c$ coincides with that of $-\bar{F}$. Using this null method, declination D and inclination I of the geomagnetic field $\bar{F}$ are measured as the direction of the helmholtz coil. The AC signal is amplified by a low-noise amplifier which employ operational amplifier units. Output signal of the amplifier is displayed on a synchroscope. Watching the synchroscope figure, the observer adjusts the direction of the helmholtz coil by means of two tangent screws of the theodolite, one is for azimuth and the other is for inclination. The DI–72 system can determine the declination D and inclination I of the magnetic field with an absolute accuracy of 1″ (Yanagihara *et al.*, 1973).

4. Computer System

All digital data obtained by the KASMMER system is processed by the data acquisition system which is a computer system with various accessories (Figure 4). The system consists of two mini-computers which are model–E600 made by HITACHI manufactory in Japan. Their computer-memories are 128 K and 256 K words, respectively. Computer 1 (128 Kword) is used for the routine processes, which are, for

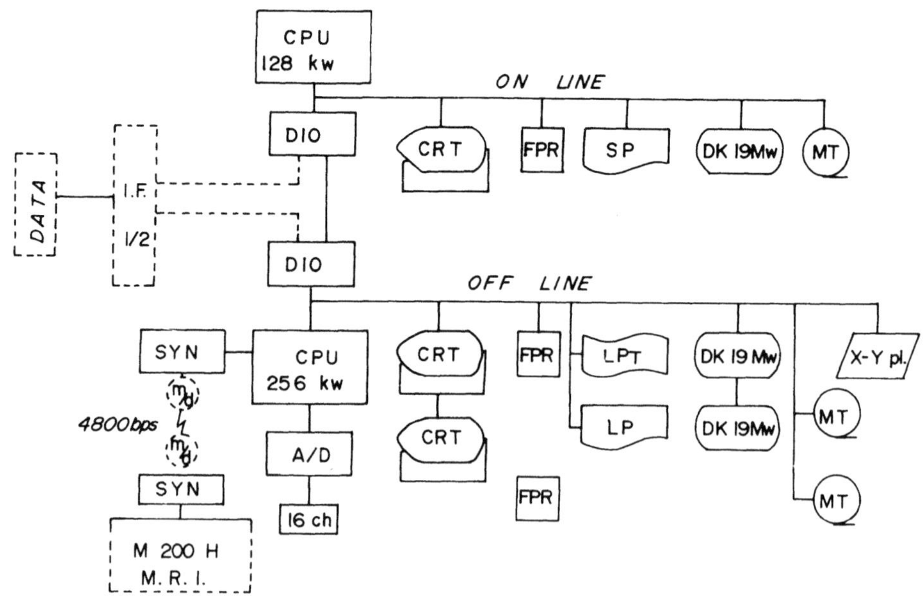

Fig. 4. Block diagram of the data acquisition system.

example, continuous aquisition of secondary data from optical pumping magnetometer, several calculations for the check of the observed data and converting the data from the computer memory to the disk memory. The BCD data of the optical pumping magnetometer are sent to the computer 1 on-line at every one-second interval. These secondary data are processed in the computer 1 according to the following processes.

(1) Judged from the apparent rate of change in secondary value and very abnormal values are rejected.

(2) Corrections of the instrumental parameters and the location differences which are given by the calibration system and stored in the computer memory, are added to the raw data of F, H, Z, Hy.

(3) Other components, D, I, X and Y are calculated from the corrected values of F, H, Z, Hy, as well as a stability factor A by the following relations,

$$D = \cos^{-1}(Hy/H) + D_0 - 60°;$$
$$I = \cos^{-1}(H/F);$$
$$X = H * \cos D;$$
$$Y = H * \sin D;$$
$$A = F - (H^2 + Z^2)^{\frac{1}{2}},$$

where D_0 is a constant determined by calibration.

(4) If any one of those calculated data exceeds a given limit, or A-value exceeds a given limit, a special mark is attached to the value.

SECONDARY DATA

KAKIOKA GEOMAGNETIC DATA MAI-BYO CHI LIST AUG. 8. 1983 OH OM - OH 1M

H= 29500.00 Z= 34000.00 D= 6. I= 48. X= 29500.00 Y= 3000.00 FO= 45500.00

	H	Z	D	I	X	Y	FO	H
OH OM 0S	645.75	705.05	37.413	61.346	444.54	477.17	470.45	0.824
OH OM 1S	645.75	705.05	37.426	61.346	444.53	477.29	470.45	0.824
OH OM 2S	645.75	705.05	37.426	61.346	444.53	477.29	470.45	0.824
OH OM 3S	645.65	705.05	37.420	61.356	444.43	477.22	470.45	0.891
OH OM 4S	645.75	705.05	37.426	61.346	444.53	477.29	470.45	0.824
OH OM 5S	645.75	705.05	37.426	61.346	444.53	477.29	470.45	0.824
OH OM 6S	645.75	705.05	37.426	61.346	444.53	477.29	470.45	0.824
OH OM 7S	645.85	705.05	37.433	61.336	444.62	477.36	470.45	0.762
OH OM 8S	645.75	705.15	37.426	61.352	444.53	477.29	470.55	0.844
OH OM 9S	645.85	705.15	37.420	61.336	444.63	477.24	470.45	0.684
OH OM10S	645.75	705.05	37.413	61.352	444.54	477.17	470.55	0.922
OH OM11S	645.85	705.05	37.420	61.342	444.63	477.24	470.55	0.859
OH OM12S	645.95	705.05	37.426	61.332	444.73	477.31	470.55	0.793
OH OM13S	645.95	705.05	37.426	61.332	444.73	477.31	470.55	0.793
OH OM14S	645.85	705.15	37.420	61.342	444.63	477.24	470.55	0.731
OH OM15S	645.85	705.05	37.420	61.342	444.63	477.24	470.55	0.859
OH OM16S	645.95	705.15	37.439	61.332	444.71	477.43	470.55	0.715
OH OM17S	645.95	705.15	37.426	61.339	444.73	477.31	470.65	0.816
OH OM18S	645.95	705.15	37.439	61.339	444.71	477.43	470.65	0.816
OH OM19S	645.95	705.15	37.439	61.339	444.71	477.43	470.65	0.816
OH OM20S	645.95	705.25	37.426	61.345	444.73	477.31	470.75	0.840
OH OM21S	646.05	705.15	37.446	61.336	444.80	477.49	470.75	0.848
OH OM22S	646.05	705.15	37.433	61.336	444.82	477.38	470.75	0.848
OH OM23S	646.05	705.15	37.433	61.336	444.82	477.38	470.75	0.848
OH OM24S	646.05	705.15	37.433	61.336	444.82	477.38	470.75	0.848
OH OM25S	646.05	705.15	37.433	61.336	444.82	477.38	470.75	0.848
OH OM26S	646.05	705.15	37.433	61.342	444.82	477.38	470.85	0.749
OH OM27S	646.15	705.15	37.439	61.326	444.91	477.45	470.75	0.785
OH OM28S	646.05	705.25	37.420	61.336	444.83	477.27	470.75	0.777
OH OM29S	646.15	705.15	37.426	61.332	444.92	477.33	470.85	0.887
OH OM30S	646.15	705.15	37.426	61.326	444.92	477.33	470.75	0.785

Fig. 5. Example of the secondary data of the KASMMER system, which are printed out for the interval 00h00m00s–00h00m30s UT on August 8, 1983.

(5) Minutely-mean data are calculated using 60 secondary data from 30 to 29 sec.

(6) Minutely-mean data from the fluxgate magnetometer are also sent to the computer 1.

The disk memory can store the secondary data for 8-days period and the minutely-mean data for 45-days period. The stored data can be easily printed out on the printer as shown in Figure 5. The data are also easily copied from disk to magnetic tape (1600 bpi) or floppy disk.

The computer 2 (256 Kword) is used for the backup system of the computer 1, and also used for nonperiodical data processing and research work as an examination of the various kinds of magnetometer.

5. Stability of the KASMMER System

In the present section, stability of the KASMMER system will be examined on the basis of data obtained during the 3-yr period of 1979–1981. Stability will be examined by monitoring correction values of the optical pumping magnetometer which are given by the calibration system. Figure 6 shows results for the F-component optical magnetometer. Monthly-mean correction values are plotted in the first part of Figure 6 for 3-yr period of 1979–1981. As shown in the figure, the stability of CF is very good. The

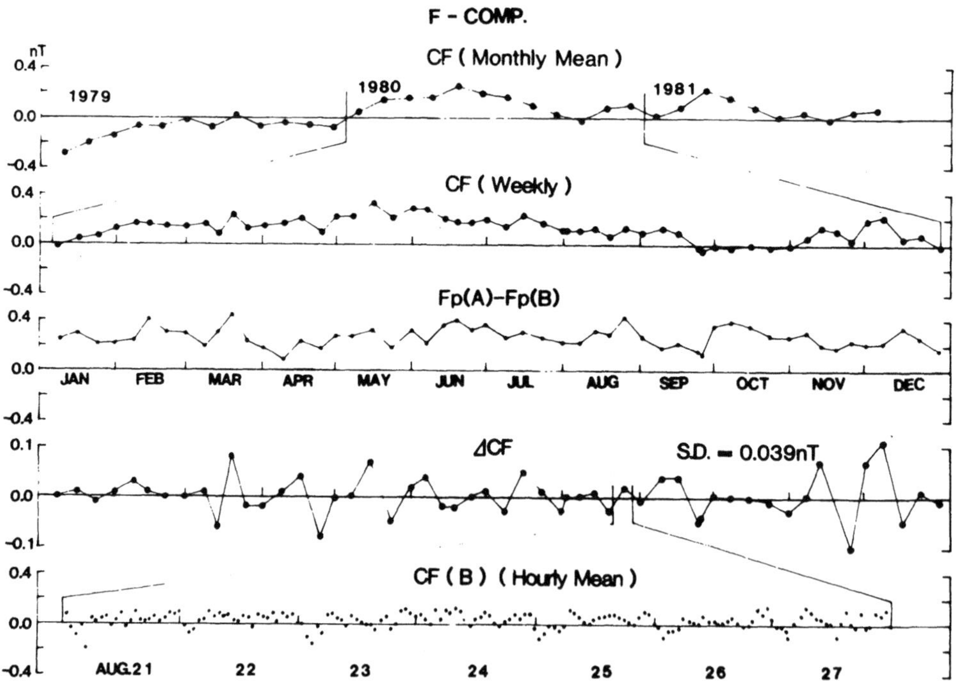

Fig. 6. Stability of the F-component optical pumping magnetometer, monthly-mean correction value (first part); weekly-mean correction value (second part); stability of the proton magnetometer (third part); variance of the weekly-mean correction value (fourth part); houry-mean correction value (fifth part).

variation of CF is less than 1 nT during 3-yr period. The F-component optical pumping magnetometer shows a satisfactory stability as a magnetic observatory system. In figure 6, we have shown weekly-mean correction value (second part), stability of the proton magnetometer (third part), variance of weekly-mean correction value (fourth part) and houry-mean correction value (fifth part). As shown in the figure, the variance of the weekly-mean correcton value is stabilized within 0.1 nT with a very small standard deviation value (0.039 nT). In conclusion, over all stability of the F-component optical pumping magnetometer is expected to be about $0.3\,nT\,yr^{-1}$. Concerning to the stability of the proton magnetometer, the precise analysis has been described by Sano (1975) and Sano *et al.* (1981).

Concerning to the H-, Z-, Hy (D)-component optical pumping magnetometers, their stabilities are less than that of F-component one. This is because of unavoidal unstability of a direction of the helmholtz coil which adds the bias field to measure a specified component of the geomagnetic field. Figure 7 shows results for the H-component optical pumping magnetometer. As shown in the figure, the monthly-mean correction value, CH, is closely correlated to the variation of the level of the helmholtz coil (third part). If a level correction is carried out for CH, a very good stability can be found as shown in the second part (CH) in Figure 7.

The above-mentioned relationship is more clearly seen for the weekly-mean

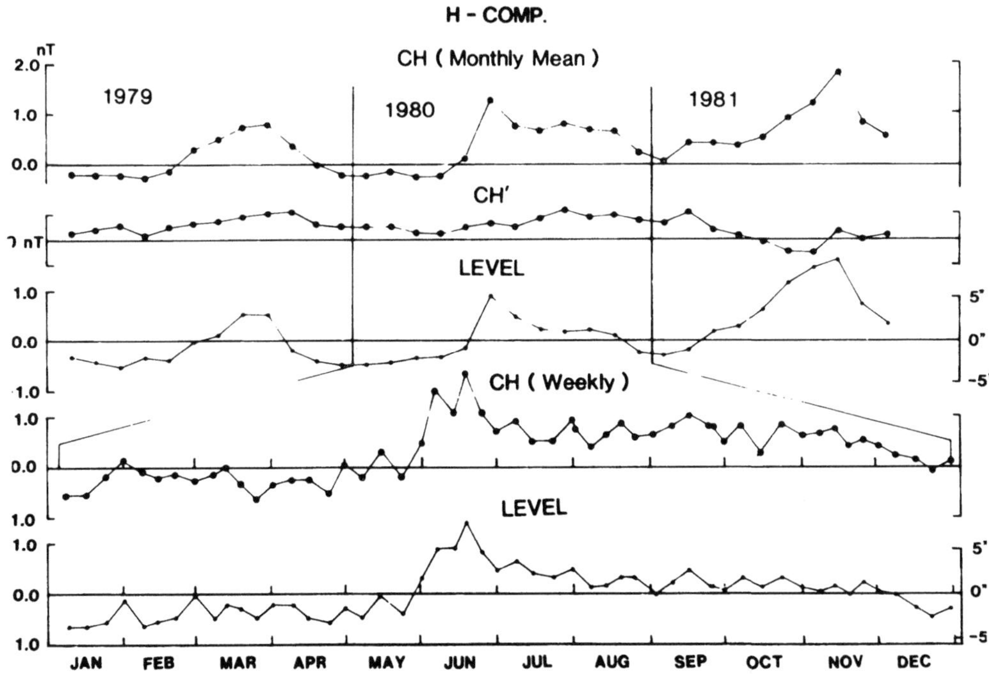

Fig. 7. Stability of the H-component optical pumping magnetometer, monthly-mean correction value (first part); monthly-mean correction value which is corrected with level of helmholtz coil (second part); monthly-mean level of the helmholtz coil (third part); weekly-mean correction value (fourth part); weekly-mean level of the helmholtz coil (fifth part); scale unit in the figure is 1 nT.

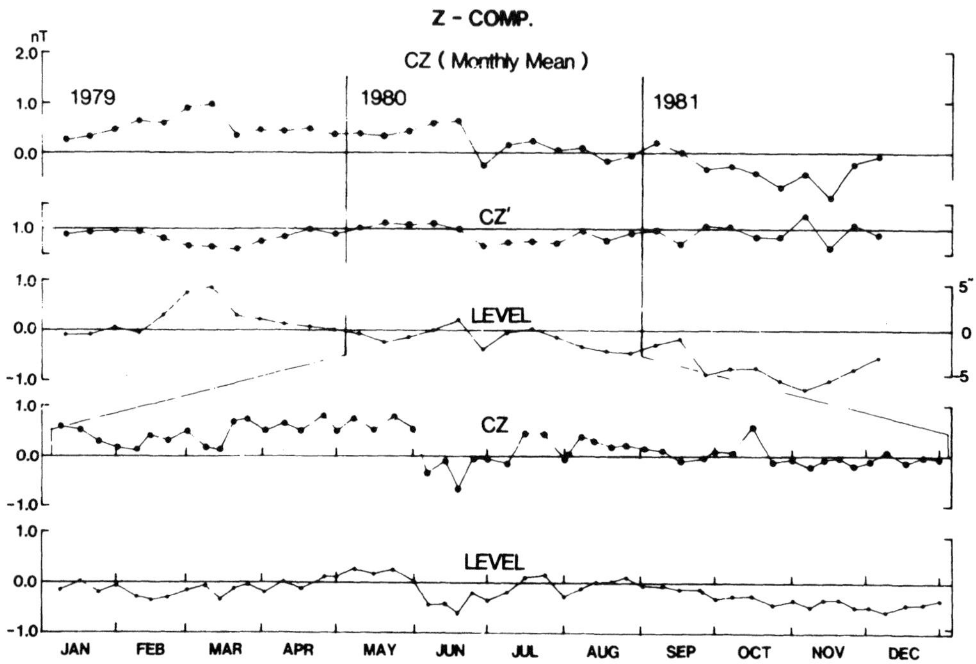

Fig. 8. Stability of the Z-component optical pumping magnetometer, monthly-mean correction value (first part); monthly-mean correction value which is corrected with level of helmholtz coil (second part); monthly-mean level of the helmholtz coil (third part); weekly-mean correction value (fourth part); weekly-mean level of the helmholtz coil (fifth part); scale unit in the figure is 1 nT.

correction value as shown in the fourth and fifth parts in the figure. The close relationship between the correction value and the coil level will be also found for the *Z*-component optical pumping magnetometer as shown in Figure 8. If correction is carried out, we could have a stability of about 0.4–0.5 nT yr^{-1} for the *H*- and *Z*-component optical pumping magnetometers.

Concerning to the *Hy* (*D*)-component optical pumping magnetometer, some complicated problems are found as shown in Figure 9, where the variation of the correction value is not caused only by the variation of the level of the helmholtz coil. Such the situation makes somewhat unstable observation of *Hy*-component comparing with the observations of other components (about 1.0 nT yr^{-1}). We must conquer the problem with Hy-component optical pumping magnetometer in the future.

6. Summary

The KASMMER system was designed more than ten years ago, 1972, to attain a stability of 0.1 nT $month^{-1}$ and a stability of 1.0 nT yr^{-1}. The KASMMER system have been improved successively in order to perform that purpose. Stabilities in the present system are 0.3 nT yr^{-1} for the *F*-component, 0.4–0.5 nT yr^{-1} for the *H*- and *Z*-

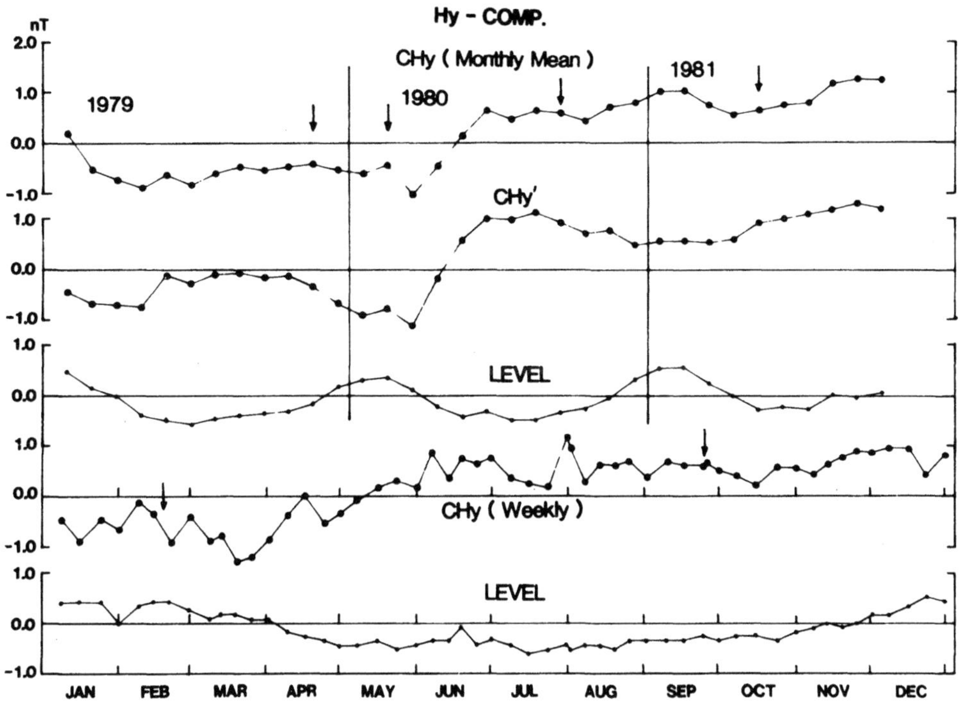

Fig. 9. Stability of the *Hy*-component optical pumping magnetometer.

components, and $1.0\,\mathrm{nT\,yr^{-1}}$ for the *Hy* (*D*)-component. So, we believe that KASMMER system has an enough ability as a standard magnetic observatory.

Acknowledgment

We would like to express our appreciation to Dr A. Harada, the director of the Kakioka magnetic observatory, Drs M. Kawamura and K. Yanagihara, the former directors of the Kakioka magnetic observatory, for their supporting the present work.

References

Yanagihara, K. *et al.*: 1973, 'New Standard Magnetic Observatory System of Kakiok (KASMMER)', *Geophys. Mag.* **36**, 217–281.

Sano, Y.: 1975, 'Stability and Accuracy of Standard Magnetic Observation by KASMMER System, *Mem. Kakioka Mag. Obs.* **16**, 121–141.

Sano, Y. *et al.*: 1981, 'On the KASMMER Magnetometer (1)', *Proceedings of the symposium on high-sensitive magnetometers*, held at Sendai in Japan, November 1981, pp. 212–219.

LONG-TERM STABILITY OF A TORQUE-BALANCE VARIOMETER WITH PHOTOELECTRIC CONVERTERS IN OBSERVATORY PRACTICE

J. JANKOWSKI, J. MARIANIUK, A. RUTA

Institute of Geophysics, Polish Academy of Sciences, 00-973 Warsaw, ul. Pasteura 3, Poland

C. SUCKSDORFF and M. KIVINEN

Finnish Meteorological Institute, Department of Geomagnetism, Box 503, SF-00101 Helsinki 10, Finland

Abstract. Conventional magnetometers of the Bobrov type were equipped with photoelectric converters and feedback loops having a feedback factor of 1000. Tests made at the Belsk and Nurmijärvi Observatories demonstrated that: (a) the long-term stability of the base-line values was the same or better than that in conventional magnetometers (e.g. at Nurmijärvi), i.e. the changes were less than 0.5 nT a month; (b) the short-term stability was the same as in the conventional magnetometers; (c) the reliability of the instrument was good, the time between failures in the instruments tested being more than a year; (d) the same instruments can be used simultaneously for normal recording and for recording magnetic pulsations.

1. Introduction

Torque-balance variometers of the classical type have been in operation at magnetic observations for over a hundred years. These instruments are extremely reliable, and easy to service and operate. Recordings of the horizontal components have a high stability with which other types of instruments have so far been unable to compete.

The main deficiency of a conventional magnetograph based on this principle is its recording system: most of the recorders are photographic. The deflection of the light beam reflected from a mirror fixed to the sensing magnet is recorded on photographic paper. But modern uses of data require electric signals from the magnetographs that would enable both analog chart recorders and digital loggers to be employed.

In this paper we will show that the above shortcoming can be remedied by using photoelectric converters with torque-balance variometers, without adding long-term instability to the system. In addition, certain technical parameters of the instrument, such as the frequency-recording band, linearity, and temperature coefficients, can be substantially improved. Our report is based on tests performed at the Belsk and Nurmijärvi Observatories.

2. The Torsion-Photoelectric Magnetometer

Brunelli *et al.* (1960) described a magnetometer based on torque-balance variometers with photoelectric converters. Since then several other photoelectric converters have been written up in the literature. The majority of magnetometers based on this principle have been designed for magneto-telluric studies in which long-term stability is not as

Geophysical Surveys **6** (1984) 367-380. 0046-5763/84/0064-0367$02.10.

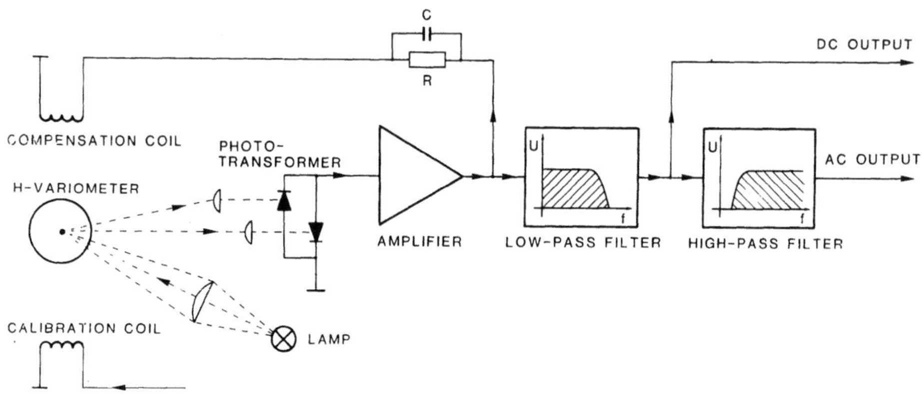

Fig. 1. Principles of the Torsion-Photoelectric Magnetometer.

important as in observatory practice. An attempt to construct a magnetometer suitable for both observatory and field use has been made at the Belsk Observatory. The first descriptions of this magnetometer were published in 1977 and 1978 (Marianiuk, 1977, Marianiuk *et al.*, 1978). Our paper describes tests on this type of magnetometer, which we will call the Torsion Photoelectric Magnetometer (TPM).

The principles of the TPM are shown in Figure 1. In this case, quartz variometers constructed by Bobrov were used as sensors (Bobrov, 1962), but any kind of conventional magnetometer would have been equally suitable. Bobrov quartz variometers are of a type in which quartz fibres are attached to both sides of the magnet so that the magnet is suspended in the fibre. The same kind of sensor can be used for recording any component of the magnetic field. During the recording the magnetic field perpendicular to the sensing magnet is kept almost at zero. Most of the field is compensated using magnets with near-zero temperature coefficients. These magnets are fixed to the quartz frame of the variometer. Additional small magnets can be used outside the variometer to produce compensation in fields different from those for which the magnetometer is designed. To these variometers, photoelectric converters are added. The electronics and photoelectric converters are similar for all three components. In a portable TPM all three sensors with photoelectric converters are in the same casing. The interior of this part of the instrument is shown in Figure 2. In addition to the three quartz variometers and the three photoelectric converters, the casing contains electronic equipment to compensate the temperature effects and record the temperature. The second part of the TPM is the box with the rest of the electronics. This unit contains equipment for compensating whatever remains of the permanent magnetic field, for selecting the sensitivity and for calibration, and the low-pass and high-pass filters for the DC and AC outputs. The output signals have ranges of ± 10 V. DC and AC – i.e. normal variations and pulsation and the temperature – can be recorded simultaneously. Figure 4 shows the console and the sensor electronics box of the TPM.

Fig. 2. The interior of the sensor component of the TPM.

2.1. Transfer function of the TPM

This paper reports the result of our tests, but it is difficult to discuss these without a theoretical analysis. For those interested, we present the main points of the theory in the Appendix.

Of practical value is the last formula in the Appendix, which shows that the output voltage $U(t)$ is a linear function of the variation of the magnetic field $\delta H(t)$. The coefficient of this linear relation is R_s/G_s, which means that, with a strong negative feedback the conversion depends in practice only on G_s and R_s, where G_s is the feedback coil factor (nT/mA) and R_s the resistance of the feedback loop. For example, the output voltage is non-sensitive of the quartz element. To be precise, it depends slightly on these parameters, because a strong negative feedback decreases the effect of a possible change by a factor of β, β being the feedback coefficient. In the case of the TPM, β is about 10^3. It should be stressed that the feedback does not correct instability in the compensation of the permanent magnetic field, or tilting of the variometers, because these parameters are not included in the feedback loop. So the long-term stability of the TPM depends

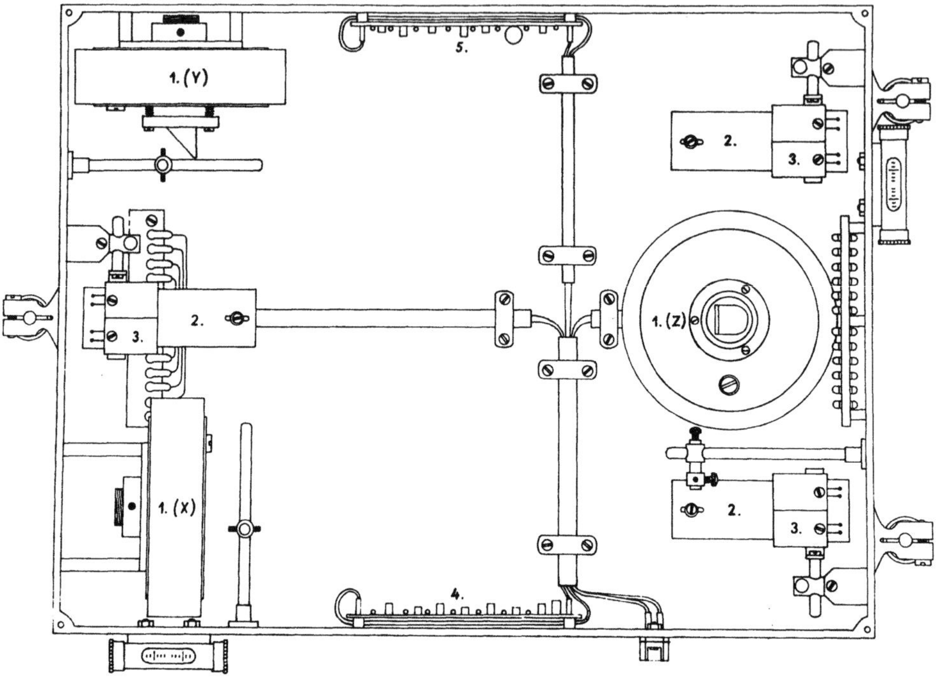

Fig. 3. Sensor component of the TPM, containing three perpendicular magnetometers (1), three lamps (2), three phototransformers (3), preamplifier (4), and temperature card (5).

mainly on changes in the compensation of the permanent magnetic field and on the angle at which the variometers are tilted. Furthermore, the feedback shifts the resonance frequency of the system to a higher frequency (see formula for ω_s in Appendix). In practice, the shift is approximately 20 Hz.

3. Tests of the Technical Parameters of the TPMs

Three Torsion-Photoelectric Magnetometers have been tested at the Belsk and Nurmijärvi Observatories. One of the instruments was constructed with three separate sensors, the second had the three sensors in a single light weight portable casing, the third instrument had a heavier and sturdier casing. The differences of construction have had an effect on the long-term stability.

3.1. Sensitivity and Resolution of the TPM

It is not easy to determine the maximum sensitivity of the TPM. The noise of the AC output has been estimated at 10^{-3} nT peak-to-peak. The sensitivity and noise were tested by running two TPMs with digital recordings simultaneously. Calculations of the variances of the two digitally recorded data sets show that the noise level of the TPM is well below 0.01 nT when recording short-period variations (the AC output in Figure 1).

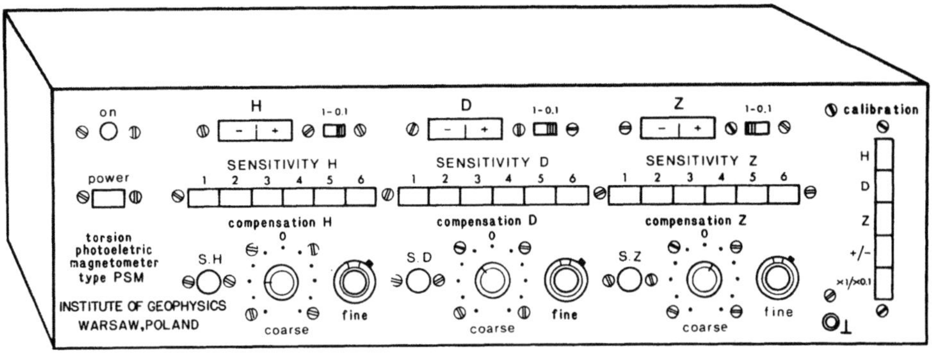

Fig. 4. The console of the TPM. This box contains most of the electronics, which are battery-operated (10 ... 15 V). The power consumption of the entire TPM is 15 W.

3.2. Linearity and Interaction between the Sensors of the TPM

Linearity was tested by varying the amplitude of the calibration signal and by measuring the output voltage. We found that the linearity of the TPM is better than 0.1%.

Owing to the strong negative feedback (see Section 2.1), the sensing magnet in the TPM does not move in practice, so interactions caused by deflexions of the sensing magnets can be disregarded. The main reason for interactions between the sensors is the effect of the magnetic field produced by the feedback coil of one sensor on the other two sensors. The interactions of the components were tested by sending a strong calibration signal to one component and measuring its effect on the others. This effect was found to be smaller than $\frac{1}{300}$ of the calibration field.

3.3. Temperature Coefficients of the TPM

The temperature coefficients of Bobrov variometers vary from 0 to 2 nT per °C, when the permanent magnetic field is compensated by magnets fixed to a quartz frame. When additional magnets on metallic rods are used, the temperature coefficients are usually larger. As a rule, the higher the value of magnetic field intensity to be compensated, the bigger the temperature coefficient. The temperature coefficient of the vertical component is greatest in middle and high geomagnetic latitudes. Temperature coefficients can be reduced with suitable electronic compensation. In practice we have found it easy to cut them down to 0.2 nT per °C in TPMs.

3.4. Effect of Tilting on the TPM

In a rigid system of sensors the effect of tilting is easy to calculate. The approximate formulas are:

$$\Delta H = Z \sin \psi$$
$$\Delta D = Z \sin \varphi$$
$$\Delta Z = H \sin \psi$$

where ΔH, ΔD, and ΔZ are the changes of the horizontal intensity, declination (in nT) and vertical component of the magnetic field, ψ is the tilting angle of the sensors in the N–S plane, and φ is the tilting angle of the sensors in the E–W plane.

The TPM is not a rigid system, so the directions of the magnetic moments of the sensing magnets do not change in an exact function of the tilting of the whole casing. The sensing magnets are suspended on fibres, and the centre of mass of the system does not always coincide with the axis of rotation. The sensitivity to tilting therefore depends also on the location of the centre of mass in the magnet. The fibres by which the sensing magnets are suspended are elastic, so the two fibres on both sides of the magnet are not in line. The deviation of the fibres from a straight line depends on many factors, mainly on the strain imposed on them. For these reasons it is not feasible to calculate the exact sensitivity to tilting of a variometer, so we determined it empirically. The results for the magnetometers tested are given in Table I. It can be seen that the values vary for each

TABLE I

Effect of tilting of the sensors by 1′ on the values recorded at Belsk; $H = 19\,050$ nT, $Z = 45\,300$ nT, and at Nurmijärvi; $H = 15\,200$ nT, $Z = 50\,000$ nT

Component	Plane in which the sensor is tilted	Rigid system	TPM 1 at Belsk	TPM 2 at Belsk	TPM 3 at Nurmi-järvi
H/X	N–S	−13.2 nT	−3.1 nT	−2.7	−3.0 nT
	E–W	0	−4.0	+3.8	3.0
D/Y	N–S	0	+3.5	+3.4	5.0
	E–W	−13.2	−1.7	−3.4	−3.5
Z	N–S	+ 5.5	+5.8	+5.0	+5.5
	E–W	0	+3.5	−4.2	4.5

magnetometer though, on the whole, the horizontal variometers with suspended magnets are less sensitive to tilting than those with rigid sensors.

Nevertheless the TPM is clearly sensitive to tilting. To achieve high long-term stability, it is therefore necessary to use stable pillars, and fit the sensors in sturdy boxes.

4. Long-Term Stability of the TPM

The final check in the suitability of the TPM as an observatory recording instrument is the long-term stability of its base-line values. Test measurements were made with two different TPMs at Belsk and one at Nurmijärvi.

4.1. Stability Tests at Belsk

Analog recording systems were used for both the TPMs tested at Belsk (51°50.2′ N, 20°47.5′ E). The base-line values were determined with vector-proton magnetometer (Marianiuk and Oberc, 1976). The measurements were made almost daily.

Due to some non-linearity in the analog recorders, the scatter of data is a little higher than for photographic recordings. Nevertheless the data contain enough material to determine the long-term stability of the instrument.

The first test was made from November 1977 to January 1978, and the results have been published (Marianiuk *et al.*, 1980). The stability was good, with a drift smaller than 0.5 nT per month.

The second test was performed using a TPM with three sensors inside a light weight casing. This test lasted one year and was made in 1982. The base-line values for the horizontal components were stable: the change was 1 min of arc per year in the D component, and 3 nT yr^{-1} in the H component.

In the vertical component the drift was much larger. It is difficult to say whether this drift was due to the weakness of the casing or to other factor such as instability of the compensating magnets. The results of the determinations are shown in Figure 5.

On comparing the results with the base-value determinations for conventional systems at Belsk, published in every Year book of the Belsk Observatory, we can conclude that adding phototransformers to torque-balance variometers does not affect the original long-term stability of the variometers.

4.2. Stability Tests at Nurmijärvi

The Nurmijärvi Geophysical Observatory (60°30.5′ N, 24°39.3′ E) has a well stabilized set of La Cour variometers, 30 yr old, with photographic recording (20 mm hr^{-1}) and sensitivities of 8.04 nT mm^{-1} in X, 3.97 nT mm^{-1} in Y and 7.04 nT mm^{-1} in Z. Base-line values are determined by Serson's method with a vector proton magnetometer (Kivinen, 1982) for H and Z, and three different declinometers for D. The absolute level of the different instruments is checked annually in Nordic comparisons of magnetic instruments. These measurements have shown the H and F standards at Nurmijärvi to be the same as the Danish (and other Nordic) standards within the accuracy of the measurements, which has been better than 0.5 nT in almost every comparison. The absolute accuracy of D, which was also included in the Nordic comparisons in 1982 and 1983, was found to be better than 0.5′, and probably better than 0.3′ at Nurmijärvi, corresponding to 2 nT (1.2 nT) in Y at that observatory.

The accuracy of the Nurmijärvi base lines can be assumed to be the same as the above, i.e. 0.5 nT in X and Z and about 1.5 nT in Y constantly, thanks to the stability of the old variometers and to the absolute measurements, made once every 10 days.

One TPM in a sturdy casing was installed in the Nurmijärvi variometer house on a pillar formerly occupied by a La Cour Quick-Run magnetograph in Summer 1982. The house is temperature controlled (18 ± 0.2 °C), so the temperature compensation of the TPM was turned off. The TPM electronics room also has a constant temperature; tests have shown a temperature dependence of 0.5 nT $°C^{-1}$ in the TPM electronics for Z. The other components have not shown any temperature effect due to the electronics. Several changes were made to the data logging system, etc., in 1982, so only data for 1983 will be used for checking on the long-term stability of the TPM.

The Nurmijärvi La Cour and TPM values have been compared as follows:

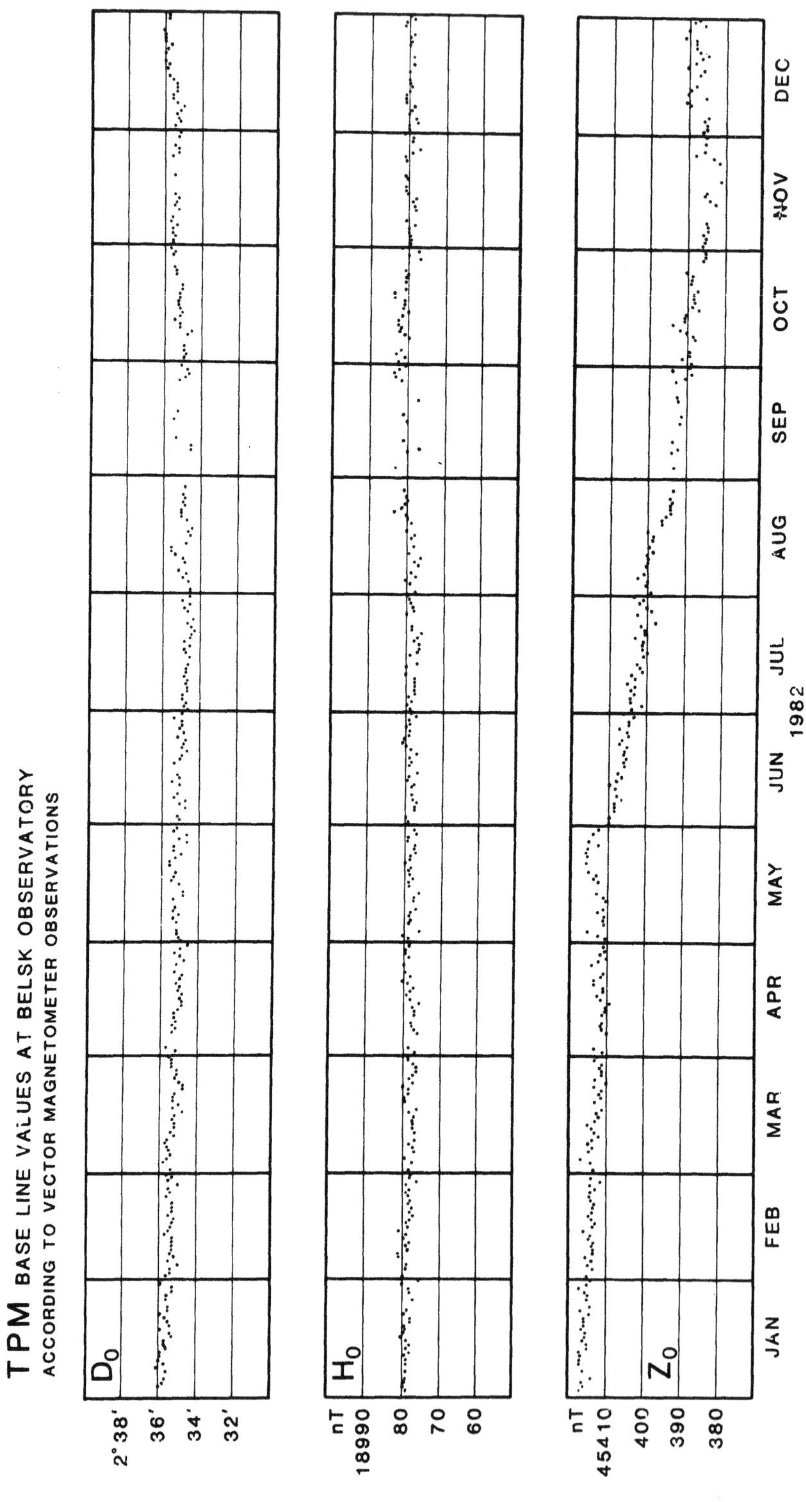

Fig. 5. Base-line values for a portable TPM, according to tests made at Belsk. Points are absolute measurements made with vector proton magnetometer.

Hourly mean values from the La Cour magnetograms have been carefully hand-scaled and calculated with the aid of the final base-line values. The corresponding hourly mean values for the TPM have been calculated from its one-minute mean values, but using the same constant base-line values for the TPM for the whole period of the test run, i.e. base line values giving the same value for the TPM as for the La Cour in October 1982.

Differences between the 'absolute' La Cour hourly mean values and the corresponding TPM hourly mean values have been then calculated for every hour since the beginning of 1983. Daily mean values have also been calculated from these differences. Figure 6 shows the daily means of the differences for the period 1 January to 31 October, 1983. Examples of the hourly differences for a magnetically quiet period lasting three days are shown in Figure 7. From Figures 6 and 7 it can be seen that the variation of the TPM level has been roughly -1 nT a month in X, -0.2 nT a month in Y and $+0.3$ nT a month in Z. During the last five months the variations have been

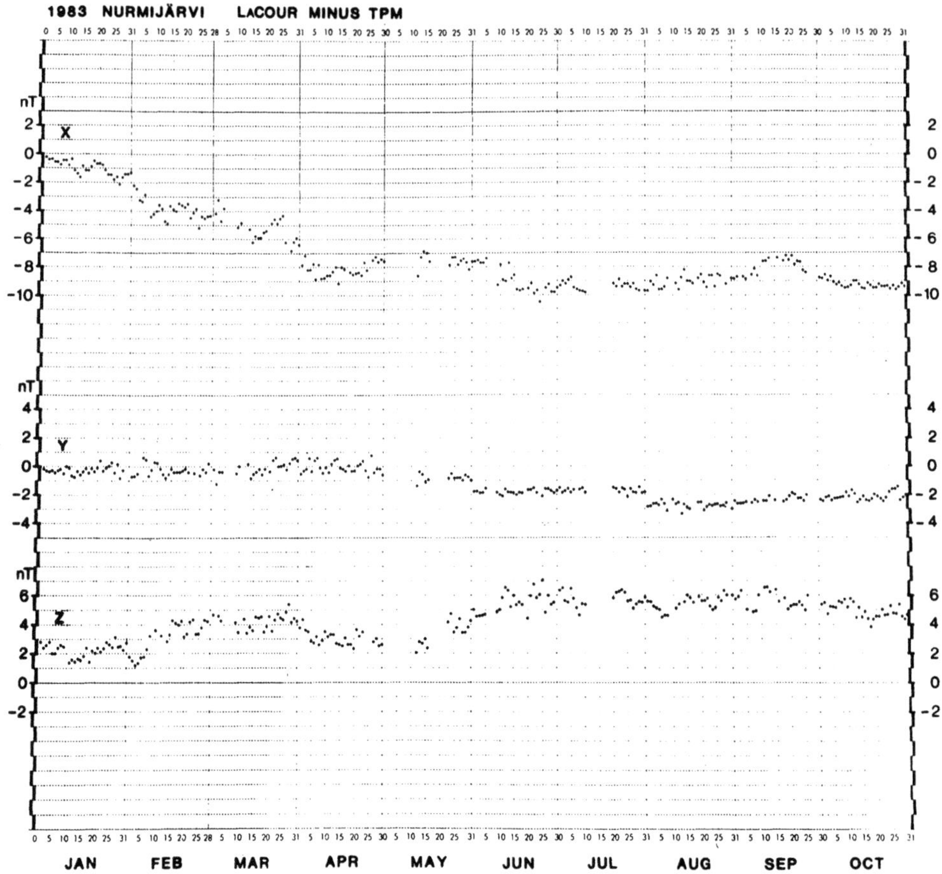

Fig. 6. Daily differences between Nurmijärvi La Cour values and corresponding TPM values, using constant base-line values for the TPM.

about 0.2 nT a month in all the components. The La Cour base line values have varied in steps of 1 nT. These steps are clearly visible in the curve for Y. The scatter of the points corresponds roughly to the scaling accuracy of the hourly mean values from the magnetograms. Hand-scaling is not accurate enough for this kind of comparison especially during disturbed periods. Periods of 1 or 2 nT higher or lower difference values lasting several days can also be detected, but are difficult to explain. The scatter in Figure 7, showing the hourly differences (c. ± 1 nT), is larger than the scatter in the daily values, the rms being 0.5 nT. These values correspond to 0.05 ... 0.1 mm on the magnetograms and can be explained by errors in the hand-scaling of the magnetograms. Even the error in the scaling glass is found to have exceeded 0.1 mm at some readings.

All the afore-going indicates that the short-term stability of the TPM at Nurmijärvi is at least as good as is that of the La Cour magnetograph used as a control, i.e. better than

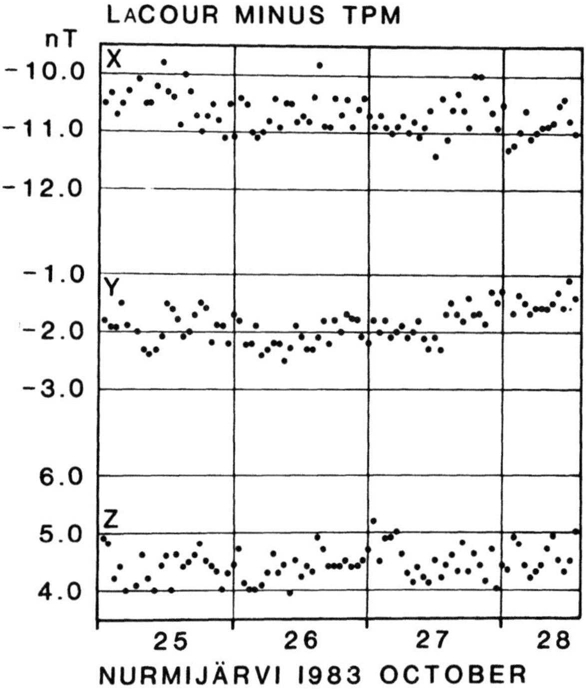

Fig. 7. Hourly differences (La Cour minus TPM) for three quiet days in October 1983.

say 1 ... 2 nT within a few days. The long-term stability is better than 0.5 nT a month in all the components.

5. Experience with the Reliability and Operation of the TPM

Seeing that the TPM is more complex than conventional magnetic variometers, it is open to suspicion of being less reliable. In principle it is true that the more sophisticated the instrument is, the bigger are the problems with its maintenance. Experience has shown that the difference in this case is not substantial. Thanks to the comparative simplicity of its electronics, the TPM itself is easier to operate than its recording system, expecially if this is a chart recorder. On the other hand, the use of a modern chart recorder eliminates the need for photographic developing, so that the total amount of work needed for the daily operation of a TPM is smaller in practice than that for conventional magnetographs. Adding a data logger and storage units like tape recorders to the TPM naturally makes it more complex.

During over a year's operation at the Nurmijärvi Observatory, no failures have occurred in the TPM or its electronics. The only failures have been in the data storage system, which is not part of the TPM. The much longer experience with the TPMs in Belsk shows that failures of the TPM or its electronics are rare.

Thanks to the high dynamic range of the TPM, it is not necessary to use separate sets of variometers to record normal magnetic variations and pulsations; the same TPM can be used for both. For safety's sake, however, it is best to run two sets of instruments, say a conventional photographic magnetograph and a TPM – to ensure that no data is lost.

Table II compares the parameters of a conventional variometer and a TPM.

TABLE II

Properties of the TPM compared to a typical conventional variometer

Type of instrument	Resolution	Linearity	Transfer function	Type of recording	Dynamic range	Temperature coefficient
Conventional	0.3 nT	$\sim \tan \Theta$	flat 1 Hz – DC	Analog photographic	~60 dB	0 – 2 nT deg^{-1}
TPM	0.01 nT	~0.1%	flat 20 Hz – DC	Analog and digital	~90 dB	0 – 0.2 nT deg^{-1}

8. Conclusions

Tests and comparisons made at the Belsk and Nurmijärvi Observatories have led to the following conclusions on the properties of the TPM (Torsion Photoelectric Magnetometer), according to the results given above:

(1) The long-term stability of a TPM is as good as or better than the stability of a well stabilized conventional magnetometer with photographic recording.

(2) The short-term stability, say from hour to hour and day to day, also seems to be as good as that of the classical magnetometers used as controls in the tests. In fact, the scatter found in the comparisons is just as likely to be due to imperfections in the magnetogram scaling glass or in the optics of the conventional magnetometer, as in the

TPM. In any case the scatter of the values was small: less than 2 nT, and usually less than 1 nT, which is the limit of accuracy in handscaling.

(3) The magnets in the TPM remain almost constant in position. This leads to linear dependence on variations of the components of the magnetic field, which is not the case in conventional magnetometers during strong disturbances.

(4) The TPM can be used for a frequency range of 0 to 20 Hz, covering all the frequencies normally needed for geomagnetic research and recorded at magnetic observatories. The only problem is to find the right recording systems for the different frequencies.

(5) Thanks to the fairly simple electronics, the time between failures in the TPM seems to be several years, and is probably almost the same as in flux-gate magnetometers. Our experience is that failures in the recording equipment and data loggers have been the main problems in the use of TPMs.

(6) The output of the TPM is ± 10 V, which makes both analog and digital recording easy.

(7) The TPM is sensitive to tilting, but not in the same way as a rigid instrument like the flux-gate. Table I shows that the effect of tilting on the horizontal components is smaller than in a rigid variometer. Nevertheless, a TPM needs a sturdy stable pillar.

(8) The TPM includes electronic compensation for the effects of temperature. Only preliminary tests of the efficiency of this compensation have been made. In most cases, auxiliary magnets have to be used to compensate for the rest of the field (unless a constant current is used for that purpose). The temperature effect of these auxiliary magnets is rather difficult to compensate. So a temperature-controlled room or box is preferable in observatory use. This is also desirable for the electronic equipment, if possible, because it, too, has been found to have some temperature effects.

Appendix

Theory behind the TPM (for further details see Marianiuk, 1977).

The governing differential equations and transmittances of the elements of the feedback loop are given in Figure 8.

The transmittance of a chain of elements connected in series equals the product of the transmittances of all the elements of the chain. The transfer function of a converter with an open feedback loop is:

$$K'(i\omega) = K_1 \cdot K_2 = \frac{M}{K} \frac{k}{\tau(i\omega)^3 + (2\varepsilon\tau + 1)(i\omega)^2 + (2\varepsilon + \tau\omega_0^2)i\omega + \omega_0^2}.$$

The transfer function of a converter with negative feedback is

$$K_s(i\omega) = \frac{K'}{1 + K'K_3} = \frac{A}{a_0(i\omega)^3 + a_1(i\omega)^2 + a_2(i\omega) + a_3} = U_s(\omega)e^{-i\phi_s(\omega)};$$

where

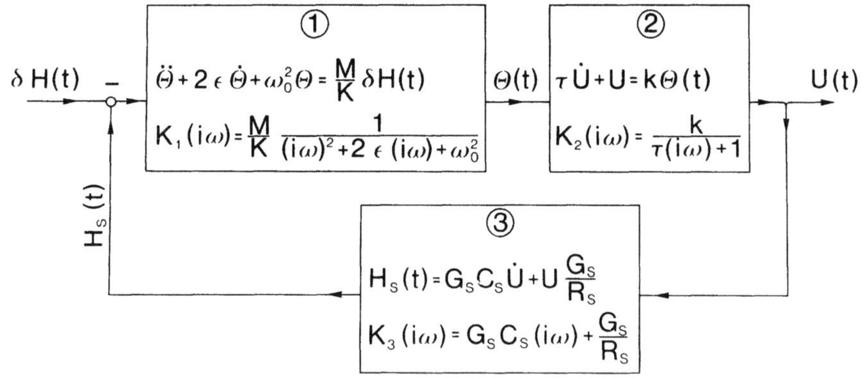

Fig. 8. Working principles of the TPM. (1) is the torsion variometer, (2) is the photoelectric converter and amplifier, (3) contains the elements of the feedback loop. $\delta H(t)$ is the variation of the magnetic field to be recorded. $U(t)$ is the corresponding output voltage. $H_s(t)$ is the field produced by the coil around the magnet which compensates the effect of $\delta H(t)$. M and K are the magnetic and inertia moments of the magnet, Θ is the angle of deflection of the magnet, ε the damping coefficient, ω_0 the natural frequency of the magnet, τ the time constant of the amplifier, k the amplification (V/rad), G_s the feedback coil factor (nT mA^{-1}), R_s and C_s are the resistance and capacitance of the feedback loop (R and C in Figure 1) and ω is the frequency of the magnet.

$$A = \frac{M}{K}k; \quad a_0 = \tau; \quad a_1 = 2\varepsilon\tau + 1; \quad a_2 = 2\varepsilon + \tau\omega_0^2 + AG_sC_s;$$

$$a_3 = A\frac{G_s}{R_s} + \omega_0^2$$

and

$$U_s(\omega) = \frac{A}{\sqrt{(a_3 - a_1\omega^2)^2 + (a_0\omega^3 - a_2\omega)^2}}.$$

$U_s(\omega)$ is the coefficient of transformation of the converter which changes $\delta H(t)$ into the output voltage $U(t)$. When $\varepsilon\tau << 1$

$$U_s(\omega) = \frac{A}{\sqrt{(\omega_1^2 - \omega^2)^2 + 4\varepsilon_1^2\omega^2}}$$

where $\omega_1 = \sqrt{\omega_0^2 + AG_s/R_s}$ is the resonance frequency of the torsion photoelectric converter, $\varepsilon_1 = \varepsilon + 1/2\,AG_sC_s$ the damping coefficient of the converter , and ε_p the conversion coefficient of δH to U.

The formula when A is high and $\omega < 3\,\omega_1$ is:

$$U(t) \simeq \frac{Mk}{K\omega_1^2}\delta H(t)$$

where

$$\omega_1^2 = \frac{MG_s}{KR_s}k = A\frac{G_s}{R_s},$$

$$\varepsilon_p = \frac{U(t)}{\delta H(t)}\left[\frac{\mathrm{mV}}{\mathrm{nT}}\right] = \frac{R_s}{G_s}\left[k\Omega/\frac{\mathrm{nT}}{\mathrm{mA}}\right].$$

References

Bobrov, V. N.: 1962, 'Seriya kvartsevych magnitnykh variometrov', *Geomag. i Aeronom.* **2.2**, 348–356.

Brunelli, E., Raspopov, O. M., and Yanovskiy B. M.: 1960, 'Vysokochuvstvitelnaya varyatsionnaya stantsiya dlya registratsii korotkoperiodnykh kolebaniy magnitnogo polya Zemli', *Sb. po geofiz. priborstr.* **5**.

Kivinen, M.: 1982, 'Magnetic results from Nurmijörvi Geophysical Observatory 1981', No. **23**, Finn. Met. Inst., 34 pp.

Marianiuk, J. and Oberc, P.: 1976, 'Measurements of *H*, *Z*, *F*, and *D* with a proton Magnetometer at the Belsk Geophysical Observatory, *Publ. Inst. Geophys. Pol. Acad. Sc.* **88**, 59–85.

Marianiuk, J.: 1977, 'Photoelectric Converter for Recording the Geomagnetic Field Elements: Construction and Principle of Operation', *Publ. Inst. Geophys. Pol. Acad Sc.* **C-4/114**/, 58–73.

Marianiuk, J., Gnoinski, A., and Szymański A.: 1978, 'Recording of Geomagnetic Field Elements with a Photoelectric Magnetometer', *Publ. Inst. Geophys. Pol. Acad. Sc.*, **C-5** /**125**/, 57–76.

Marianiuk, J., Jankowski, J., and Sucksdorff, Ch.: 1980, Combining Classical Magnetometers with Simple Electronics, a Promising Way of Modernizing a Magnetic Observatory', *IAGA News* **19**, 109–111.

EXPERIENCE OF GEOMAGNETIC FIELD RECORDING WITH A FLUXGATE MAGNETOMETER HAVING A BRIDGE SENSOR

TORSTEN BERGMARK

Sveriges Geologiska, Undersokning Geological Survey of Sweden, Uppsala

Abstract. A fluxgate magnetometer has been developed at the Geological Survey of Sweden. It measures the field in three orthogonal directions and has sensor elements forming bridges. The instrument can have direct readout in both digital and analogue form with a resolution of 0.1 nT.

An instrument of this type has been adopted for stationary recording of geomagnetic elements in digital form. Special attention has been paid to insure good long term stability and high reliability. Experience from routine recording is described.

1. Introduction

For some time the magnetic observatory group of the Geological Survey of Sweden has wanted to introduce digital three-component magnetic measurements at their observatories. During this time the airborne measurements section of the Geological Survey has been developing a three-(*tri*)-component fluxgate magnetometer (TRIX). The main goals of this development work have been to increase recording speed, resolution and accuracy of measurements and to perform three-component air-borne magnetic measurements with support of an inertial navigation system. The final goal has been to improve geophysical and geological interpretation of airborne measurements.

The airborne magnetometer incorporated many attractive properties for observatory measurements. Therefore the observatory group started to develop a digital observatory version of the instrument after obtaining a prototype of the airborne instrument. Thereby the fluxgate instrument was modified, special features were added in order to facilitate observatory work and a recording system was constructed.

The following sections will describe the observatory magnetometer system and experience with its use in recording work. The airborne instrument is now available from the Swedish Geological Company. This company was formed in 1982 when the airborne measurements section and other parts were split off from the Geological Survey.

2. Fluxgate Magnetometer

The three-component magnetometer is based on the well-known fluxgate measurement principle. However its sensor is of a special construction that deserves description.

The basic unit of the sensor consists of a mu-metal foil placed in a cylindrical support for a solenoid. The solenoid has a radius of 2 mm and a length of 40 mm. For each magnetic component, four of these units are placed parallel in a plexiglass cube as can

Geophysical Surveys **6** (1984) 381-391. 0046-5763/84/0064-0381$01.65.

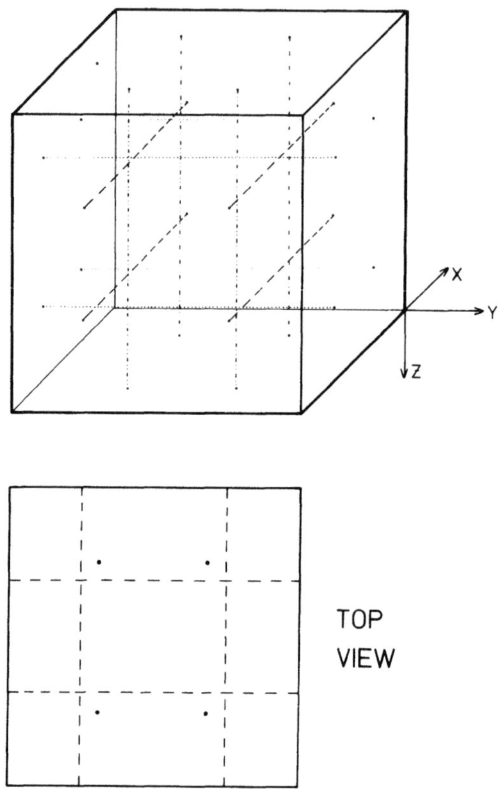

Fig. 1. Schematic drawing of the TRIX sensor, indicating the mechanical arrangement of the solenoids.

be seen in Figure 1. With this mechanical arrangement a high degree of symmetry is obtained in the three orthogonal measurement directions along the cube axes. The mechanical alignments of the sensor elements along the measurement directions is better than 5×10^{-2} degrees. The four basic units of a component are electrically connected to form a bridge which is schematically shown in the block diagram of Figure 2. Upon the sensor bridge a square voltage wave is driven with a frequency of 20 kHz. The bridge configuration gives a high output signal from the sensor in combination with a rapidly decreasing field from the drive current. After amplification, the bridge response is detected in a synchronous detector and a comparator determines if the response shall result in a step up or down of a 16-bit binary counter. The counter output is fed to a digital-to-analog converter. After amplification the converter output current is used as a compensation current in the sensor bridge. The counter output is also the measurement output. The three components are in principle identical.

The electronic circuits are basically the same as in the airborne magnetometer. Detailed descriptions are available from the Swedish Geological Company.

The modifications of the airborne magnetometer are aimed to increase its stability

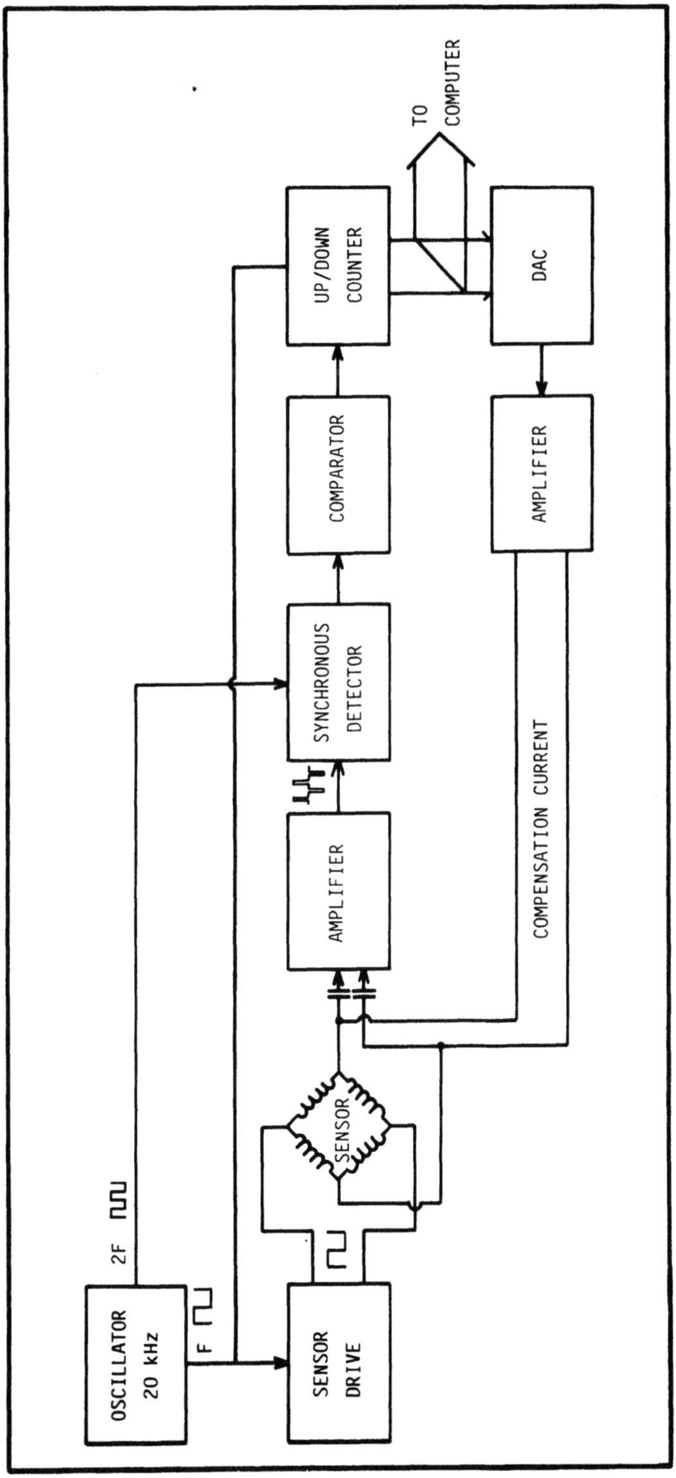

Fig. 2. Block diagram of the electronic circuits of the TRIX magnetometer.

and resolution. It was especially necessary to improve the long term stability of the instrument while measurement speed and range could be reduced.

Since the present instrument is intended for recordings of slow variations similar to those made by La Cour variometers it has not been found necessary to make a measurement of its frequency response. However the airborne magnetometer has a tracing speed of more than 7×10^4 nT/s. In the present instrument the response for high frequencies is reduced by a RC-filter with a time constant of 10^{-3} sec.

In order to suit Swedish conditions the following measurement ranges were choosen for the magnetic elements X, Y, and Z:

X	0	– 32 000 nT	0.5 nT;
Y	– 16 000	– 16 000 nT	0.5 nT;
Z	0	– 64 000 nT	1.0 nT.

The last column shows the resolution corresponding to one least significant bit of the counter outputs. By introduction of a controlled oscillation in the servo loop and mean value calculation in the following microprocessor, the resolution could be further increased at the expense of the recording speed.

The stability was to some extent improved by the choice of more stable electronic components and better regulation of some critical voltages. However the main causes of drifts in the measurements are temperature changes of the sensor and the digital-to-analog converter. These drifts have been reduced by stabilization of the temperature of these parts. They were placed in boxes of styrofoam with a wall thickness of at least 5 cm. The temperatures in the boxes were regulated to about 40 °C by electrical heating controlled by thermistors. The heater of the sensor box consists of a closely twisted pair of copper wires in order not to disturb the magnetic field of the sensor.

The sensor holder was also provided with screws and indicators for simple orientation. The additions to the sensor resulted in a sensor box with the approximate dimensions 30 × 40 × 40 cm, a size which would have been intolerably large in the airborne system.

3. Data Recording System

The data recording system was designed for permanent recording of 10-sec mean values from the TRIX magnetometer and to facilitate a simple and reliable operation of the system. It was decided that the system also should be able to record proton magnetometer measurements and that it should be able to expand to allow recording of measurements from other instruments which may be added in the future.

Figure 3 is a schematic diagram of the present system. An Intel microprocessor 8080 was choosen as master of the system while the use of an Intel 8741 interface processor was dedicated to the control of and recording from the TRIX and proton magnetometers.

The 8741 processor is programmed to add and count measurement values from the TRIX at full transmission speed, about 2 kHz, during a time interval determined by the

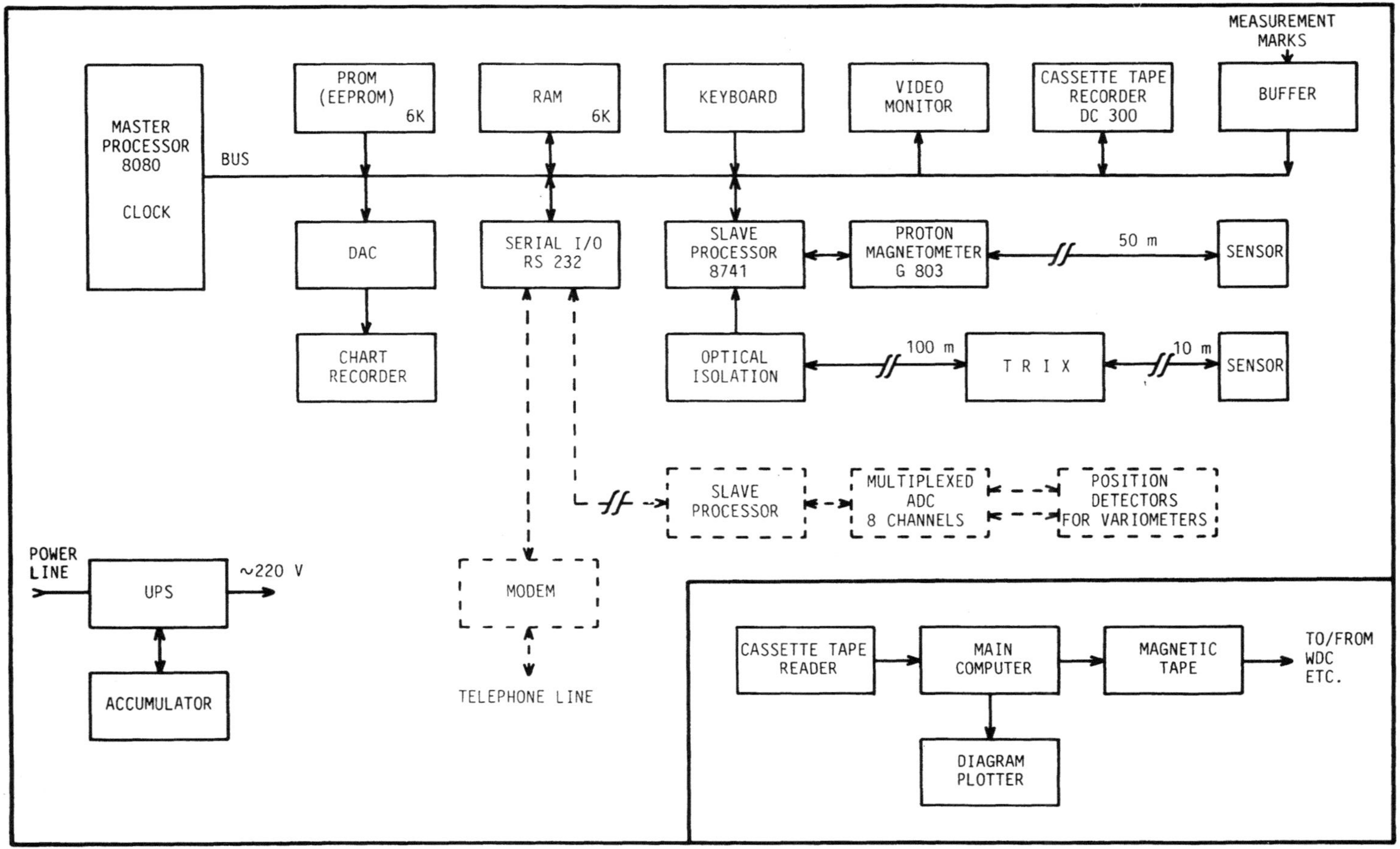

Fig. 3. Block diagram of the data recording system. The dashed parts are available but they cannot be used until the controlling software is completed.

master processor clock, at present 10 sec. During the same interval a single proton magnetometer measurement in stored in its memory. At the end of the time interval the 8741 processor transfers its measurement information to the master processor and then it normally starts a new measurement cycle. The master processor calculates mean values which then are converted to magnetic field values by use of stored calibration factors. The master processor also stores measurement values permanently on a magnetic cassette tape. The DC 300 type of cassette has been choosen because of its high reliability and high storage capacity, more than 2.5 megabytes.

The master processor has a set of programs which can be controlled by keyboard entries. The programs perform such tasks as start, control and finish of magnetometer recordings, display of measurement values and status information on a video monitor, transfer of measurement values to a chart recorder and tests of the proper function of the system. A rather valuable feature is the possibility to store comments on measurements. A serial input/output interface makes it possible to connect modems, printers and data sources such as the slave processor indicated in Figure 3. The latter processor will be used to control tests of a position sensitive light-spot detector which can have light sources such as the classical La Cour variometers.

The recording system has to be placed far from the magnetometer sensors in order that it shall not disturb the measurements with its many magnetic parts. Furthermore the sensor cable of the TRIX should be kept rather short. In our case we had to use separate buildings for the recording system and the TRIX. Due to high common mode voltages between the two buildings, an optical isolation had to be inserted in the transmission line.

The use of microprocessors in the recording system makes it sensitive to even very short power disturbances. Therefore we have added an uninterruptible power system (UPS) which is primarily fed by the power line and secondarily by accumulators.

The cassette tapes are transferred to the main computer of the Geological Survey. There the data is analysed and transfered to magnetic tapes e.g. of the type sent to the WDC. Diagrams of the recordings are also plotted on a precision plotter.

4. Results

The detector output of the TRIX magnetometer has been observed with an integration time constant of about 0.1 sec. The observed peak-to-peak value of the noise envelope was found to be less than 0.1 nT. Since the observations were made in a varying magnetic field they show that the intrinsic resolving power of the instrument is at least 0.1 nT.

Linearity measurements have shown errors of up to a few nT when some of the most significant bits of the digital-to-analog converter are changed. These errors limit the useful ranges of the instrument when higher accuracy is needed. However the useful ranges for a linearity better than 1 nT are at least 4000 nT, which is more than adequate for geomagnetic field recordings. A linearity of 0.1 nT can be obtained in ranges of about 400 nT. If these ranges are considered too small it is possible to extend them up to

30 times since there are now available digital-to-analog converters with 30 times smaller errors.

Measurements of the temperature coefficients of the sensor show that these are less than 10 nT K^{-1}. By means of the temperature regulation the sensor temperature is held constant to within 10^{-2} K d^{-1}, which is adequate for 0.1 nT resolution of the instrument. The temperature coefficients of the digital-to-analog converters are much lower, about 0.5 nT K^{-1}. Therefore the demands on the temperature regulation of the converters are easily fulfilled for a stability requirement of 0.1 nT.

We have found no detectable influence on the measurements from the electrical heating of the sensor.

We have had no facilities for measuring the instrument stability with an accuracy of 0.1 nT. However recordings from very quiet geomagnetic periods have demonstrated a short term stability of 0.1 nT for at least 10 min. The following table is an example of such a recording. Each line shows 10s-mean values of the components *X*, *Y*, and *Z*. The fourth column shows proton magnetometer values with 0.3 nT resolution.

X	Y	Z	F
15476.2	492.0	48080.0	50517.5
15476.1	492.0	48080.0	50517.5
15476.1	491.9	48080.0	50517.5
15476.1	491.9	48080.0	50517.3
15476.1	492.0	48079.9	50517.5
15476.1	492.0	48079.9	50517.5
15476.1	492.0	48080.0	50517.5
15476.1	491.9	48080.0	50517.0
15476.0	492.0	48080.0	50517.3
15476.2	492.0	48080.0	50517.3
15476.1	491.8	48080.1	50517.5
15475.9	491.8	48080.0	50517.5
15476.0	491.9	48080.0	50517.5
15475.9	491.9	48080.0	50517.5
15475.9	491.8	48080.0	50517.8
15475.9	491.9	48079.9	50517.5
15475.9	491.8	48079.8	50517.5
15476.0	491.8	48079.9	50517.8
15476.0	491.8	48079.8	50517.3
15475.8	491.9	48079.9	50517.8
15475.7	491.9	48079.7	50517.5
15475.8	491.9	48079.8	50517.3
15475.9	492.0	48079.7	50517.0
15475.9	492.0	48079.8	50517.5
15475.8	492.1	48079.8	50517.8
15475.8	492.1	48079.9	50517.3
15475.8	492.0	48079.9	50517.5
15475.7	492.0	48079.9	50517.5
15475.7	492.0	48079.9	50517.5
15475.7	492.0	48079.9	50517.0

The stability over longer terms has been evaluated by comparisons with classical variometer recordings which have been calibrated by means of absolute measurements. These comparisons suffer from the difficulty of making high resolution scalings of the classical recordings but we have not yet had the possibility to make any better evaluations.

Two different types of comparisons have been made. In the first type, TRIX recordings were plotted with scale values equal to those of the classical recordings. The diagrams were then put on top of each other and the differences between the base-line of the classical recording and a constant base-line of the TRIX were measured at each full

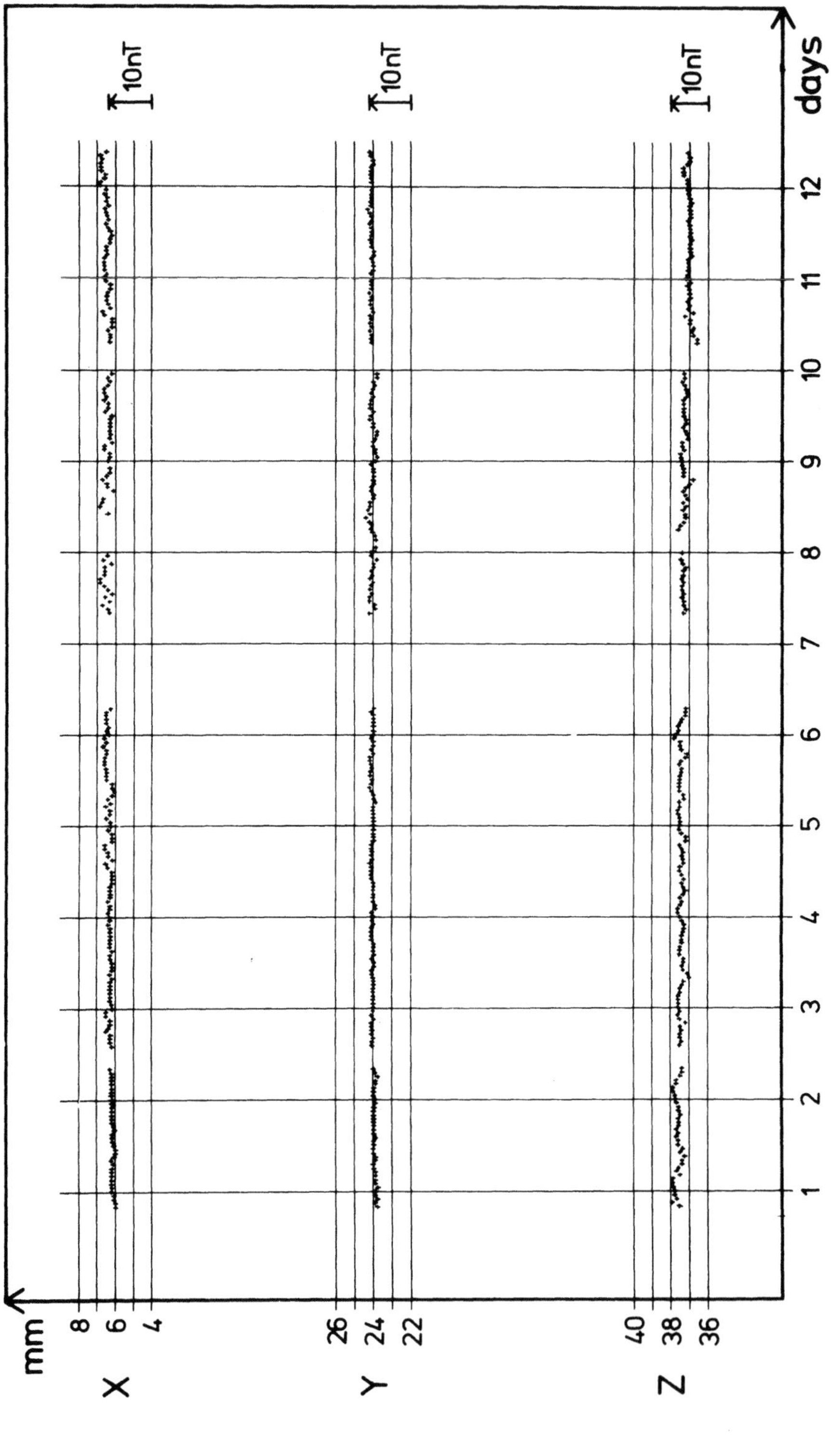

Fig. 4. Base-line differences between magnetograms from the TRIX magnetometer and the classical variometers at the Lovö observatory.

hour. Results of such a comparison are shown in Figure 4. During the 12 day long period one observes a drift of about 2 nT for the X and Z components. The superimposed faster fluctuations of about ± 1 nT seem to be partly real measurement variations for the Z component.

In the second type of comparison we have used measurement marks on the classical recordings and the corresponding 10 sec mean-values from the TRIX magnetometer. The measurement mark events are automatically stored in the TRIX recordings as indicated in Figure 3. The classical recordings have been scaled and field values calculated by means of the best available scale- and base-values. From the TRIX recordings, field values have been calculated by means of constant scale- and offset-values. The differences between the field values obtained in these two ways are shown in Figure 5. The vertical bars in these figure indicate the replacement of a bad power supply and the repair after a lightning damage. These repairs have apparently caused small changes in the measurements.

If one examines the X and Y components for the hole period and the Z component after 830 517 one finds a rather smooth drift. The drift is no more than 1, 2, and 1 nT/month for the X, Y, and Z components respectively. The larger scatter for the Z component before 830 517 is probably due to errors in the comparison. These errors

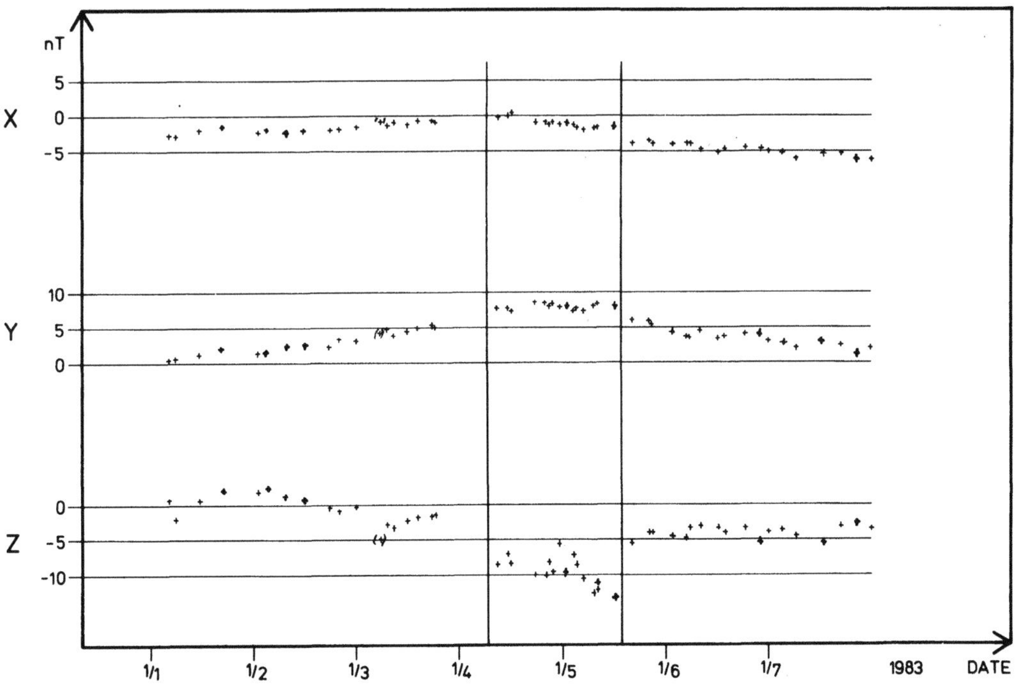

Fig. 5. Differences between field values obtained from the TRIX magnetometer and the classical variometers at the Lovö observatory. Vertical bars indicate repairs of the TRIX.

might be due to rather short artificial disturbances which are most prominent in the Z component.

Measurements have shown that the three measured components are orthogonal to each other to within ± 0.2 degrees which is about the accuracy of these measurements.

Our demands on the reliability of an observatory recording system caused us to install the UPS. Before its installation we had quite a number of power interruptions resulting in sometimes rather long data drop outs. The UPS has only been in use for a short time; therefore our experience of its use is limited. However, we have found that its rectifier for maintenance charging of the accumulators is equipped with switched diodes which feed transients back to the power line. These transients are picked up by our Elsec proton magnetometers and disturbs their measurements. These magnetometers have solenoidal coils in their sensors. Provided with this experience we found that the same type of disturbance also emanates from some types of variable DC power supplies.

We have also had two interruptions of the recordings due to malfunctions of electronic parts as mentioned above and indicated in Figure 5.

Mainly because of the interruptions due to the missing UPS, we have got recordings from only about 85% of our first half year of operation. We estimate that this figure would have been 98% if we had had the full system in operation.

5. Conclusions and Prospects

It is our experience that the present magnetometer measurement system is well suited for digital observatory recordings. A resolution and accuracy of about 1 nT should be obtainable if calibrations are made in intervals of 1–2 weeks. Short term variations can be recorded with a resolution of 0.1 nT.

We except that our digital system will make most manual evaluations of optical recordings unnecessary, although we will still keep the optical recording as a reserve system.

The time resolution of the system can easily be improved to at least 1 sec. However a continuous recording with such a resolution would yield unmanageably large amounts of data. The amount can be considerably reduced if only certain events are recorded with the higher resolution. A selection of events can be made by means of on line analysis with a microprocessor. The computerized observatory makes it possible to introduce new features of observatory recording, such as transfer of data and status information by telephone calls and automatic transfer of data, error warnings and storm alerts.

Finally we will give an illustrative demonstration of the large ranges of the TRIX magnetometer in Figure 6. This figure shows the recording of the X-component during an extremely violent storm which started 1982-07-13, 16:18 UT, when the instrument happened to be under test. The minimum value is about 5039 nT below the normal value. It would be interesting to know if anyone has recorded a larger deviation. At the same event, we found the following maximum deviation for Y, Z, and D: 2509 nT,

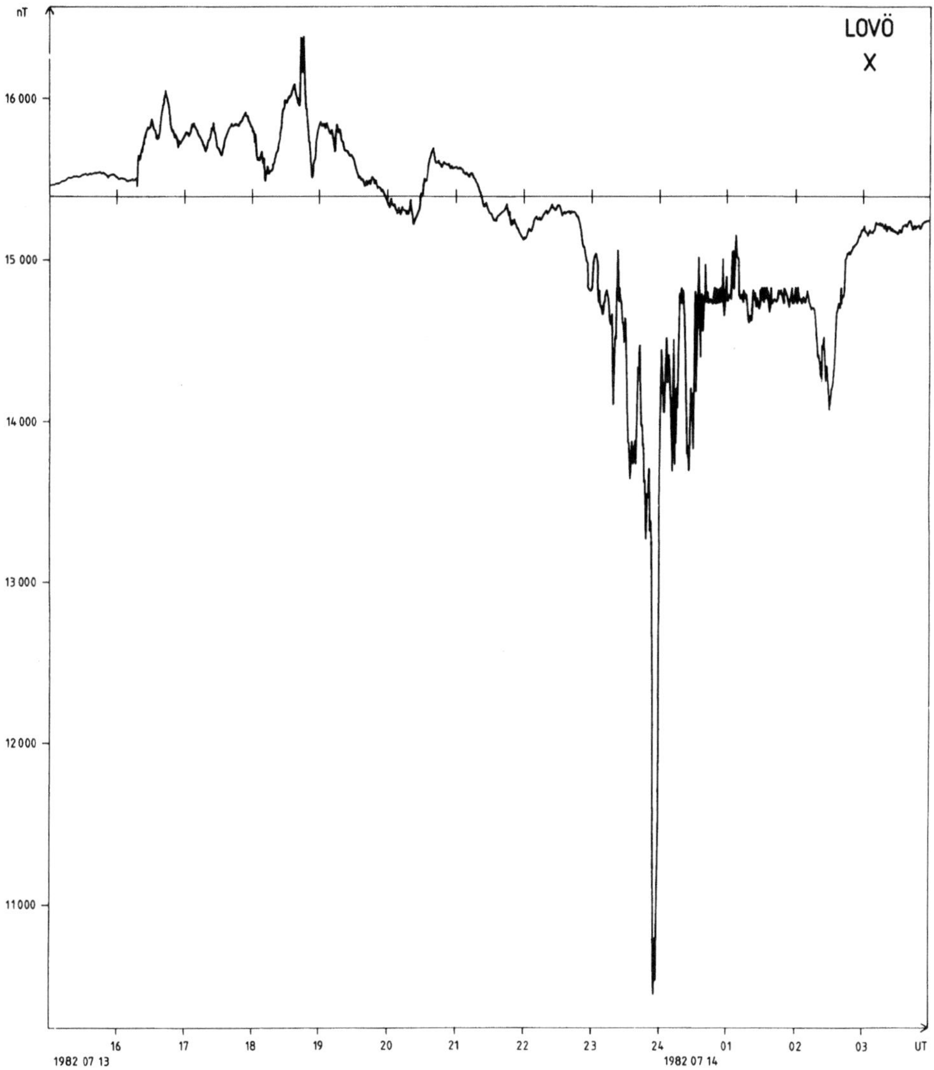

Fig. 6. TRIX recording of the X component at the Lovö observatory during the extremely violent storm starting on 1982-07-13.

818 nT, 11°. The maximum rate of changes were 467, 212, and 199 nT per 10 sec for the X, Y, and Z components respectively.

THE DIGITAL RECORDING SYSTEM OPERATED AT THE U.K. MAGNETIC OBSERVATORIES

A. J. FORBES and J. C. RIDDICK

Geomagnetism Research Group, British Geological Survey, Edinburgh EH9 3LA

Abstract. A digital recording system has been operated at the U.K. observatories since 1979. The magnetic elements *H*, *D*, and *Z* are monitored continuously by a triaxial fluxgate magnetometer operating in the variometer mode. Ordinate values are recorded at 30-sec intervals. Total field and the temperature of the variometer chamber are measured at the start of each 10-min data block.

Fluxgate magnetometer baseline values are derived from absolute measurements made by proton vector magnetometer and Kew declinometer in the conventional manner.

In this paper the data handling procedures are described and the quality of the data over a period of two years is discussed.

Introduction

The Geomagnetism Research Group (GRG) of the British Geological Survey (formerly the Institute of Geological Sciences) operates three magnetic observatories in the U.K. at Lerwick, Eskdalemuir and Hartland.

TABLE I

U.K. Magnetic Observatories

Observatory	Geographic coordinates		Geomagnetic coordinates		Lower limit for $K = 9$
Lerwick	60° 8′ N	358°49′ E	62°.5	88°.6	1000 nT
Eskdalemuir	55°19′ N	356°48′ E	58°.5	82°.9	750 nT
Hartland	50°59′.7 N	355°31′ E	54°.6	79°.0	500 nT

A digital recording system has been operated, on an experimental basis, at the observatories since 1979. The magnetic elements *H*, *D*, and *Z*, monitored by a fluxgate magnetometer, are recorded at half-minute intervals using a low power data logger. Ancillary values of *F*, measured by proton magnetometer, and temperature *T* are recorded at 10 min intervals. The recording medium, a certified data cassette, has the capacity to sustain 12 days unattended operation if the need arises.

Replay facilities for transcribing the cassette data onto disk, and processing the data have been developed for the GRG data laboratory at Edinburgh. Cassettes are normally returned to the laboratory from the observatories at weekly intervals. Data handling programs for producing minute mean component values, hourly values, and computer plotted magnetograms are in routine use.

Preliminary monthly data books containing hourly values, *K* indices, and reduced

Geophysical Surveys **6** (1984) 393-405. 0046-5763/84/0064-0393$01.95.

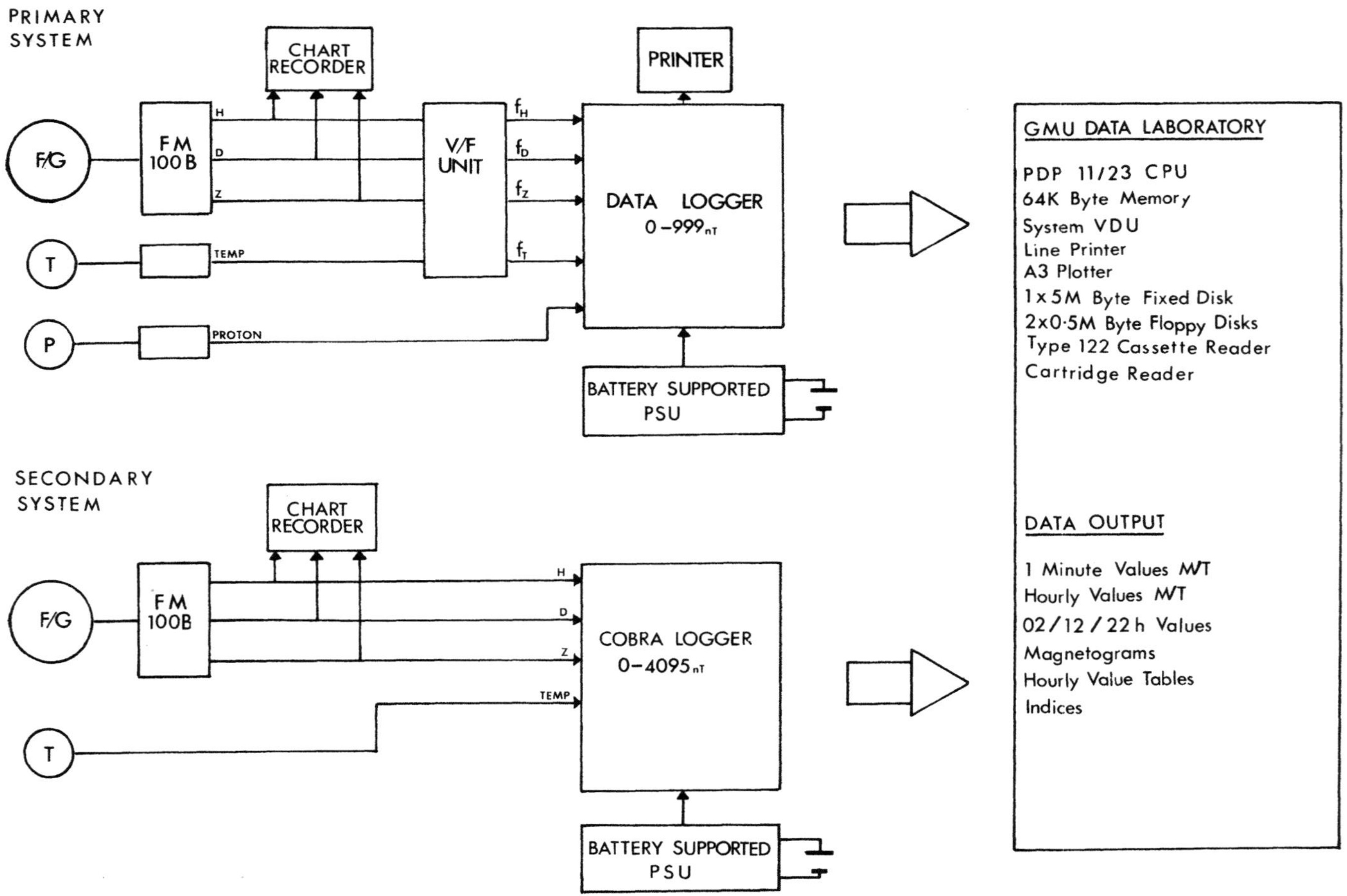

Fig. 1. The digital recording systems at the U.K. magnetic observatories. Data cassettes recorded on the loggers are processed in the GRG data laboratory.

scale magnetograms can be produced within 10 days of the end of the month during which the data were recorded.

Fluxgate magnetometer baseline values are derived in the conventional way from the absolute measurements of H, D, Z, and F made at the observatories. The system performance is monitored by comparing the digital data with proton magnetometer and La Cour magnetograph values.

2. Magnetometer Recording System

Two separate recording systems are operated at each observatory. These are referred to in Figure 1 as the 'primary' and 'secondary' systems. The magnetic components H, D, and Z are monitored using an EDA type FM 100 B fluxgate magnetometer. This is essentially an electronic variometer which measures the variation of a field component relative to a baseline value. The baseline value for the H and Z components is established by a manually adjustable offset field, which cancels the vector component directed along the magnetic axis of the fluxgate sensor element, at the time of setting up the magnetometer. Baseline stability therefore is dependent on the constancy of this offset field and the stability of level of the fluxgate sensing head. In the particular case of the D fluxgate element, which is aligned normal to the magnetic meridian when the magnetometer is set up, zero offset is applied.

The output voltages from the magnetometer ($\pm$ 10 v equals $\pm$1000 nT) are converted, in the primary system, into three proportional frequencies f_H, f_D, f_Z by a four channel voltage to frequency (V/F) conversion unit. The fourth V/F channel is utilised to derive a frequency f_T proportional to the temperature of the environment at the fluxgate sensing head. Knowledge of this temperature permits correction of the measured fluxgate ordinate values to compensate temperature variations, using predetermined magnetometer temperature coefficients. It has been found that both the magnetometer electronics and sensing head are susceptible to temperature variations, which argues for maintaining the magnetometer in a temperature controlled environment. Since November 1982 the magnetometers have been established in the variometer house at each observatory.

Voltage to frequency conversion was adopted because it was intended to operate the data loggers in observatory accommodation at distances of up to 300 m from the magnetometers. This practice was discontinued when it was found that lightning induced voltages in the long lines caused damage to the data loggers, with a consequent loss of recorded data.

A chart record of the H, D, and Z component variations, equivalent to a real time magnetogram, which can be used for scaling approximate values of the K index, is available from a 3 channel potentiometric recorder connected into the magnetometer, or alternatively into the data logger. At Lerwick, where the paper chart record is advanced by a stepping motor drive, the chart speed is adjusted to the conventional magnetograph 20 mm/hr^{-1} with rapid advance at midnight to separate the individual day length records.

Hour	Proton	T	H	D	Z
2305	0019	4800	1589	2601	3617
1589	2600	3619	1590	2600	3621
1592	2597	3621	1591	2595	3620
1594	2590	3618	1597	2584	3615
1601	2578	3612	1605	2576	3609
1603	2578	3605	1604	2577	3600
1607	2571	3595	1609	2565	3587
1609	2568	3581	1605	2574	3574
1601	2578	3568	1597	2586	3559
1594	2590	3557	1592	2583	3561
1589	2575	3560			
9365 (Day)	9962	4801	1585	2576	3559

Fig. 2. Format of a data block recorded on the observatory data logging system.

The data logger, (Riddick *et al.*, 1981) counts the V/F frequencies at 5 sec intervals for 1 sec. All four frequencies are counted simultaneously. At the end of this period the count information is latched into store. At 30 sec intervals the currently stored H, D, and Z counts are written sequentially to tape. On tape the data are organised into 10 min blocks comprising 63 data words. Each block contains an hour marker, a proton magnetometer F value and a temperature value, followed by 20 sets of half-minute values of H, D, and Z. At midnight 'days' information is substituted for the hour mark. Figure 2 shows the final 10 min data block of day 364, recorded during the period 2350–2359.5 hr, and the first six data words of the initial block of day 365. The H, D, Z, and T samples are prefixed with channel identifiers 1–4; the day number is prefixed with a 9.

The data logger is constructed round a low power incremental cassette recorder (Memodyne type 201) which records in CNRZ format on two tracks. The data word length is hardwired to 16 bits. Each word is framed by 2 bit interword gaps, and each 10 min data block is similarly framed by 16 bit end of block (EOB) gaps. A 615 bit per inch packing density permits 12 full days recording on a 300 ft. length data cassette. As the type 201 recorder has no read-after-write facility the data that is written to tape cannot be confirmed and the possibility exists that an entire week's data can be lost as a consequence of operator error or an undetected data logger fault.

The proton magnetometer is polarised three seconds before the start of block under control of the data logger clock. In the logger, a proton counter card, which duplicates the six decade counter in the magnetometer electronics, counts the same stream of pulses as the magnetometer counter. On completion of the counting period the count information is latched into store, and subsequently written to tape. It was intended that the 10 min proton magnetometer F values would be used to standardise the fluxgate magnetometer using the method described by De Laurier *et al.* (1974) but the practice was discontinued when it was found that erratic proton F values were corrupting the processed H and Z values.

During the period when absolute measurements are made at the observatory, by declinometer and proton vector magnetometer (PVM) the fluxgate ordinate values are output onto the primary system printer at half-minute intervals. This allows baseline values for the fluxgate system to be calculated in the same way as for the La Cour magnetograph.

The secondary system provides insurance against data loss due to malfunction or over-ranging of the primary logger. In the secondary logger the four (H, D, Z, and T) frequency counters are replaced by a single printed circuit card containing an input analogue switch and an analogue to digital converter (ADC). The voltages output from the FM 100 B and temperature sensor are sequentially switched into the ADC at 200 mS intervals. At the end of each conversion the 12 bit binary word generated by the ADC is written to tape together with the appropriate channel identifier. This modification expands the dynamic recording range from 999 to 4095 nT and eliminates the requirement for a separate V/F conversion unit with a consequent saving in power consumption.

Both recording systems are mains powered with standby ± 24 volt lead acid batteries maintained on continuous float charge. The batteries have the capacity to sustain recording, without data loss, during mains interruptions of up to 24 hr duration. The likelihood of a power loss for this period of time is extremely low at Lerwick and Eskdalemuir as both observatories are equipped with diesel powered standby generators.

3. Data Processing

Data cassettes recorded on the observatory data loggers are returned, by mail, to the GRG data laboratory at weekly intervals. The laboratory (see Figure 1), a relatively modest facility based on a PDP11/23 minicomputer with 64 Kbyte memory, permits the transcription, editing and processing of data to create a file of minute values. Subsequent processing is carried out on the main in-house PDP 11/70 computer.

The laboratory peripheral equipment is generally self-explanatory. The system VDU has a Tektronics compatible graphics facility which permits the display and editing of data.

Data is stored on a 5 Mbyte disk. A similar disk drive on the PDP 11/70 allows for data transfer to the main computer. The A3 X-Y flat-bed plotter is used to create reduced scale magnetograms of suitable quality for the monthly preliminary data books, and standardised scale value magnetograms for K index scaling (see Riddick and Stuart, 1984).

Cassette tapes are read on a Memodyne type 122 read-only deck, which is connected into the PDP 11/23 via a DRV11 interface card. On replay the inter-word gaps, which frame each data word, generate word synch pulses which indicate the availability of a 16 bit parallel data word at the recorder output. A full cassette is read into the system in 25 min.

A cartridge reader provides the capability of reading DC-300A data cartridges

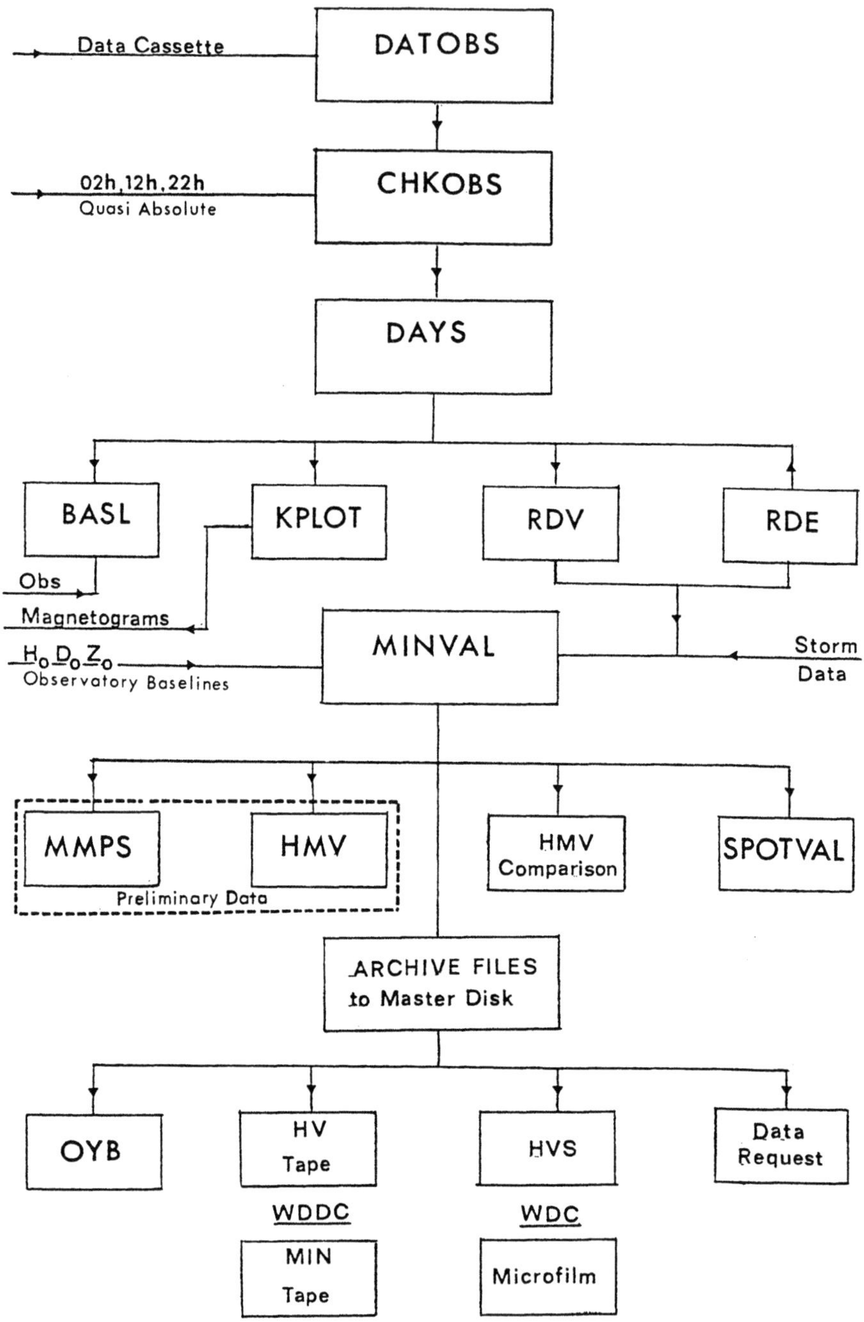

Fig. 3. Observatory digital data handling procedures.

recorded in ANSI compatible format. Experimental recording, using this medium, has been tried at Lerwick but was discontinued due to problems encountered when attempting to operate the cartridge recorder from a battery supported inverter supply.

Systems and user programs are held on floppy disk. A user friendly suite of programs has been developed for handling the observatory data. Figure 3 shows the inter-relation of the individual programs which fulfil the following functions:

DATOB
On receipt of the cassette from the observatory the recorded data, written in CNRZ format, is transcribed to a single binary format disk file. (When the data is subsequently validated the cassette is cleaned in a bulk eraser and returned to the observatory for re-use).

CHKOBS
Checks the transcribed data for timing and/or sequence errors which may have arisen during transcription, or as a result of operator error or instrument malfunction. Fluxgate ordinate values at 0200 hr, 1200 hr, and 2200 hr are extracted at this point for checking purposes. If the line printer output reveals serious problems with the 'primary' system data at this stage the 'secondary' system data is processed.

DAYS
Organises the binary data file into separate day length ASCII files to facilitate subsequent handling. Part-day files containing the data recorded at the beginning and end of a cassette are merged.

BASL
Absolute measurements of *H*, *D*, *Z*, and *F*, supplied by the observatories on a proforma which accompanies the cassette, are fed into this program, which calculates fluxgate baseline values. These values are confirmatory only as the observatories have the capability of calculating their own baseline values, and responsibility for baseline allocation.

KPLOT
The temperature corrected *H*, *D*, and *Z* fluxgate half-minute ordinate values are plotted, without baselines, to provide magnetograms, which are fed back to the observatories for *K* index scaling. It is important to establish if an accurate *K* index can be scaled from a computer generated magnetogram at this point in time as this may be the fall back position if the derivation of *K* indices by direct digital methods proves unacceptable. (All three U.K. observatories contribute to Kp, Hartland also contributes to the *AA* index).

RDV
Allows a day file of ASCII *H*, *D*, and *Z* ordinate values to be viewed on the VDU as a

magnetogram. As the observatory logger records every half-minute sample without recourse to filtering, occasional faulty data points, which would plots as transient spikes on a computer magnetogram, are inevitable. RDV allows expansion of the time scale so that individual ordinate values can be identified on the VDU screen.

RDE
Is used to edit out faulty points, identified by means of RDV, by interpolation. Artificial trace offsets of a known time duration can also be corrected.

MINVAL
Minute mean values of *H*, *D*, and *Z* are created by adding temperature corrected minute mean ordinate values to the fluxgate magnetometer baseline values allocated by the observatories, and confirmed by BASL.

MMPS
Prints out tables of hourly mean values of *H*, *D*, and *Z* for inclusion in the monthly preliminary data book. Tabular data derived from the primary and secondary systems may be compared at this stage to check the validity of the data.

SPOTVAL
Prints out spot component values at 0200 hr, 1200 hr, and 2200 hr for comparison with proton and La Cour derived values.

At the end of the month the MINVAL disk files are archived onto a master disk on the PDP 11/70. This disk can be accessed to satisfy customer requests for data and magnetogram copies. At the end of the year the magnetic tapes containing U.K. observatory hourly values and minute values are updated and sent to the World Digital Data Centres. Camera ready copy of the hourly value tables, which serve as the basis for publishing the Annual Geomagnetic Bulletin, are printed as soon as the final Quiet and Disturbed days are notified. The possibility of producing magnetograms and hourly value tables on micro-fiche, for the World Data Centres, from magnetic tape, is currently being investigated.

4. Process Control and Evaluation of the Digital Data

The data available from the observatories for comparison with the digital data are summarised in Table II.

4.1. Process control

The digital data are process controlled by:

(a) Absolute measurements; used to calculate baseline values for the magnetographs and fluxgates. The measurements are graphed together with the allocated baselines at the observatories. Inspection of the graph allows a continuous subjective appraisal of

TABLE II

Summary of the data available from the observatories for comparison with the digital data

Source of data	Frequency of data	Components	Instrument	Observatories
Absolute measurements	2–3 per week	H, Z, F	PVM	Ler, Esk, Had
Quasi-absolute measurements Output onto printer	3 per day at 0200 hr, 1200 hr, 2200 hr	Z, F H_{calc}	PVM	Ler, Esk
La Cour spot values	0200 hr daily	H, D, Z F_{calc} is checked against F_{proton}	Magnetograph	Ler, Esk, Had
La Cour hourly mean values		H, D, Z	Magnetograph	Ler, Esk, Had
Fluxgate spot values	3 per day at 0200 hr, 1200 hr, 2200 hr	D	Schonstedt Fluxgate Magnetometer	Esk
Quasi-absolute minute values logged onto DC 300 A cartridge	not continuous	F, Z H_{calc}	2 proton magnetometers	Esk

stability and performance. The observatory calculated baseline values which take account of the field variation during the observation are more accurate than the confirmatory values obtained from BASL whose program input is a mean field value centred at the mid-point of the observation.

(b) Quasi-absolute measurements are obtained by switching a pre-set current into the PVM bias coil to 'cancel' H. The resultant Z measurement is described as quasi-absolute because it is affected by such factors as mislevelling of the PVM theodolite base, coil misorientation and any residual component of the H field. The values obtained yield 'baseline' values when the relevant fluxgate ordinate values are subtracted. These values are not quantitatively accurate but they provide qualitative assurance that the digital data is consistent within an approximate ± 5 nT tolerance. At Eskdalemuir the Schonstedt fluxgate magnetometer 0200 hr value provides a check on the D baseline.

(c) La Cour 0200 hr values; all three observatories contribute to a scheme whereby participating European observatories send their 0200 hr H, D, and Z La Cour values to Wingst Observatory for comparison (Voppel, 1981). These 0200 hr values are compared on a routine basis with the fluxgate H, D, and Z values derived from the SPOTVAL program.

(d) La Cour hourly values; hourly value tables obtained from the primary and secondary digital systems make possible comparison of both systems at hourly value and daily mean level. In the event of non-agreement the La Cour values are used to identify the faulty data.

One further possibility of process control, when the system proton magnetometer is operational, is to monitor the difference $\Delta F = F_{\text{proton}} - (H_{FG^2} + Z_{FG^2} + D_{FG^2})^{1/2}$. It has been found convenient to monitor ΔF on the line printer together with the temperature in order to confirm the temperature correction. The difference ΔF is not necessarily zero as it can include a site difference component.

All the recorded data are viewed on three occasions, in the course of running RDV, KPLOT, and MMPS. The process control procedures are intended to ensure that the correct baselines and scale values are applied.

4.2. Evaluation

It can be seen from Table II that various options are possible for comparison purposes. The most accurate magnetometer available at each observatory is the proton magnetometer/PVM. At the time of making an absolute measurement a conscientious and skilled observer might expect, given quiet field conditions, to achieve a measurement accuracy of $H, Z \leqslant \ln T$; $D \leqslant 0.1$ min of arc.

Table III shows the RMS scatter of the 'measured minus the allocated' baseline values for the La Cours and the fluxgates during 1982–83. The fluxgate baselines are those obtained from BASL except where indicated otherwise. The level and morphology of the magnetic activity almost certainly influences the quality of the baseline determinations at each observatory. At Hartland where the lower limit for $K = 9$ is 500 nT, the quality of the La Cour baseline control is extremely tight and the discernible difference in the RMS scatter in favour of the La Cours is significant. This is partially attributable to the fact that BASL derived fluxgate baselines are less accurate than the

TABLE III

RMS values of the measured baseline minus allocated baseline values for the La Cours and fluxgates

Observatory	Instrument	H	D	Z	Year
Lerwick	La Cour	1.18 nT	0.26′	1.67 nT	1982
	Fluxgate	1.53	0.37	1.59	
	La Cour	1.03	0.27	1.11	1983
	Fluxgate	1.33	0.47	1.55	
Eskdalemuir	La Cour	1.54	0.28	1.05	1982
	Fluxgate	1.45	0.31	0.85	
	La Cour	1.15	0.22	1.15	1983
	Fluxgate	1.28	0.34	0.93	
Hartland	La Cour	0.91	0.11	0.86	1982
	Fluxgate	1.54	0.21	1.55	
	La Cour	0.73	0.13	0.88	1983
	Fluxgate	1.34	0.20	1.31	
	Fluxgate[a]	0.95	0.19	1.07	1983

[a] Derived from observatory calculated values.

observatory derived values, as instanced by the comparative figures for 1983 at the foot of the table. Cross talk has been observed between components at all three observatories. This effect is also most noticeable at Hartland.

The reason for the poor D baseline at Lerwick and Eskdalemuir is not understood. It may be significant that a different pattern of declinometer is used at Hartland, where the magnet is rarely detached from its suspension. It is also surprising that the fluxgate D baselines are markedly inferior to the La Cour baselines since the D element is subject to a zero offset field, which eliminates a potential source of baseline instability.

The performance of the digital system relative to the La Cours may be gauged by monitoring the difference

$$\Delta F_{LC} = F_{\text{proton}} - (H_{LC^2} + Z_{LC^2})^{1/2}$$
$$\Delta F_{FG} = F_{\text{proton}} - (H_{FG^2} + Z_{FG^2} + D_{FG^2})^{1/2}$$

at 0200 hr. The error ΔF is not necessarily zero as it may include a component due to site value difference, but it should remain sensibly constant. Observed variations in ΔF are attributable to such diverse factors as the adopted baselines and scale values, incorrectly scaled (or logged) ordinates and occasional erratic proton measurements. Figure 4 shows the monthly values of the standard deviation of the daily 0200 hr values ΔF_{FG} relative to the monthly mean of the daily values. The results confirm that the digital data appears less reliable than the La Cour data at Hartland. At the other two observatories this is not so apparent. The high values of standard deviation for the fluxgate system obtained at Lerwick during the first half of 1982 are attributable to the fact that an uncalibrated FM 100 B was in use during the period when the primary instrument was returned to EDA Instruments for repair.

5. Conclusions

The digital recording system has performed reasonably well. During 1982-83 the data loss, with both the primary and secondary system operational, is estimated at less than 1%. The most significant data losses since 1979 have been caused by lightning induced damage to the data loggers, and by a protracted mains interruption at Hartland during the winter of 1981 which discharged the then inadequate standby batteries.

The FM 100 B fluxgate magnetometer is sub-standard for observatory purposes. No attempt has so far been made to improve the instrument. Cross-talk is observed between components and there is some doubt about scale value stability. The existing digital system is regrettably deficient in calibration facilities.

The data logger, which owes its design to the mobile data logging unit which was used for micropulsation recording during the International Magnetospheric Study (IMS), 1976-79 (Riddick *et al.*, 1976) has obvious disadvantages, which can be overcome by introducing a microprocessor or minicomputer based logging system. If observatory staffing levels are to be reduced, the development of a system which will permit fully automatic operation is a pre-requisite. Work on such a system will begin in 1984.

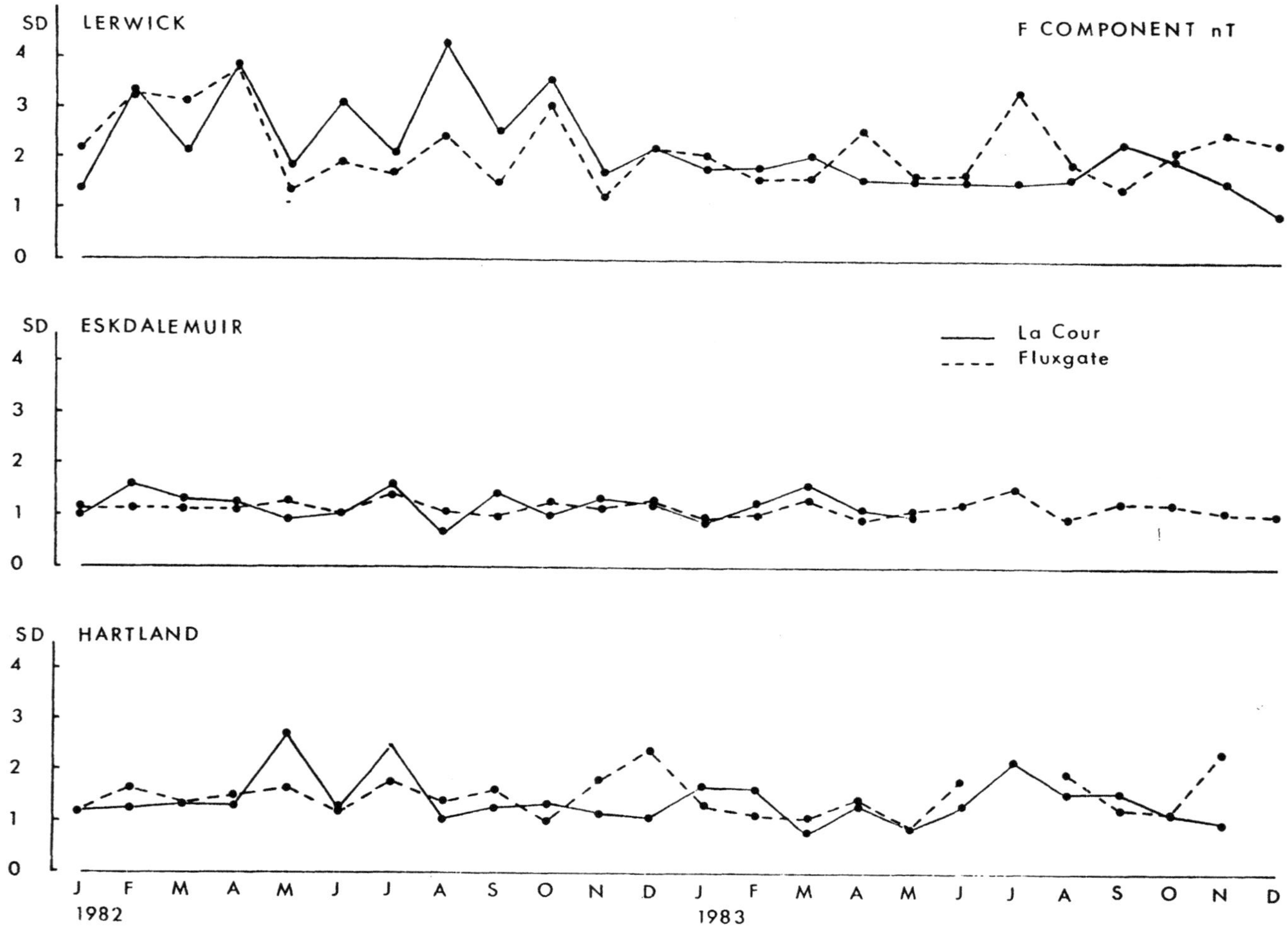

Fig. 4. A comparison of the 0200 hr calculated F values derived from the La Cour and the fluxgate data with proton magnetometer values. The figure shows the monthly standard deviation of the difference values ΔF_{LC}, ΔF_{FG} relative to the monthly mean of the differences. La Cour values are defined by the solid lines and fluxgate values by the dotted lines.

The quality of the digital data appears to be marginally inferior to that obtained from a La Cour magnetograph. One considerable advantage, however, of operating the digital system is that it has created a valuable data base of U.K. observatory one minute values that would not otherwise exist. Customer requests for data (most notably one minute F values) and records can now be met promptly with a greater degree of flexibility than was previously possible.

No fluxgate derived data has so far been published and the observatory K index values continue to be derived from La Cour records. The 1983 annual Magnetic Bulletin will be the first yearbook to contain digitally derived data.

Acknowledgements

Our thanks are extended to John Mcdonald who constructed much of the described equipment, to Dr C. A. Green who developed the editing programs RVD and RDE, and to the staff at the U.K. magnetic observatories for their patience and assistance in the course of this project.

This paper is published by permission of the Director, British Geological Survey (NERC).

References

Riddick, J. C., Brown, J., and Forbes, A. J.: 1976, 'A Low Power Moveable Observatory Unit for Magnetometer Array Application', Geomagnetism Unit Report No. 17.

Riddick, J. C., Forbes, A. J., and Green, C. A.: 1981, 'The Recording and Processing of Digital Magnetic Data from the U.K. Observatories', Geomagnetism Unit Report No. 27, 1981.

Riddick, J. C. and Stuart, W. F.: 1984, 'The Generation of K Indices from Digitally Recorded Magnetic Data', *Geophys. Surv.* **6**, 439–456. (this issue).

Voppel, D.: 1981, Contributed paper IAGA Edinburgh Meeting.

De Laurier, J. M., Loomer, E. I., Jansen Van Beck, G., and Nandi, A.: 1984, 'Editing and Evaluating Digitally Recorded Geomagnetic Components at Canadian Observatories', *Pub. of the Earth Physics Branch* **44**, No. 9.

GEOMAGNETIC MEASUREMENTS MADE ON THE MOVING ICE SHELF AT HALLEY, ANTARCTICA

D. A. SIMMONS and J. R. ROUSE
British Antarctic Survey, Natural Environment Research Council, High Cross, Madingley Road, Cambridge CB3 0ET, U.K.

Abstract. The difficulties involved in making geomagnetic measurements from the moving ice shelf at Halley are considered, as are measurements giving information on this movement. These indicate that the observatory has been moving westward, accelerating from 400 m yr^{-1} in 1969 to 800 m yr^{-1} in 1980, and rotating at up to 30 min of arc yr^{-1}. The effects of both rotation and tilt on the two types of variometers installed are examined. It is concluded that the geomagnetic data obtained from Halley since 1969 are best treated as variation data.

1. Introduction

A magnetic observatory has been operated at Halley (75.5° S, 26.9° W) on the Brunt Ice Shelf, Antarctica, since 1957. Originally set up by the Royal Society for the International Geophysical Year, as described by Brunt (1960), the operation was taken over by the British Antarctic Survey (then known as the Falkland Islands Dependencies Survey) in 1959. Whilst a conventional observatory can take for granted the mechanical stability of the buildings and the instruments placed inside them, this is not the case at Halley. The ice shelf on which the station is situated is both moving from East to West, and rotating in the horizontal plane. Rotation of the ice shelf about a horizontal axis and movement of the piers set into the ice can cause a similar rotation or tilting of the instruments. Meanwhile steady accumulation of snow over the observatory buries the buildings so that they become successively further and further below the surface.

This paper describes briefly how the problems of interpreting geomagnetic observations made at Halley have been tackled, summarises the measurements, and considers some of the results for the period from 1968–1982.

2. Location and Equipment

The location of Halley, on a floating ice shelf in the South Eastern side of the Weddell Sea, is shown in Figure 1. Figure 2 shows the successive positions of the station and an accelerating westward movement of up to 800 m yr^{-1}. By 1967 the original buildings had become uninhabitable and a new station was built a few km to the East. Since then the observatory has occupied the same site, though there has been some replacement of individual buildings crushed by the ice. A new station was built in 1983, and the instruments were moved there in February 1984.

The geomagnetic instruments in use during the period 1968–1979 were conventional observatory instruments. There were a set of normal 'La Cour' variometers and an

Geophysical Surveys **6** (1984) 407-417. 0046-5763/84/0064-0407$01.65.

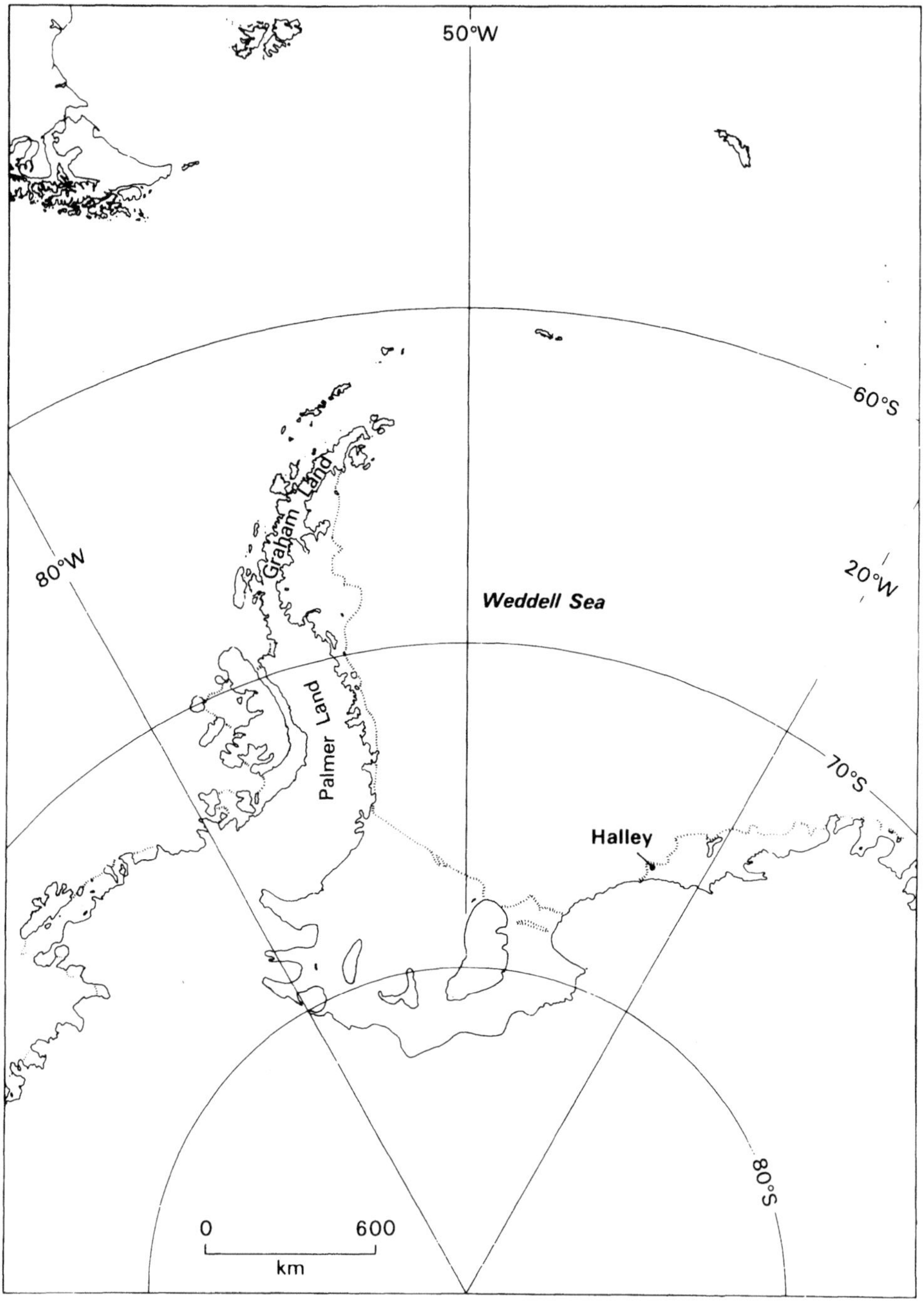

Fig. 1. Weddell Sea and surrounding area, showing the position of Halley.

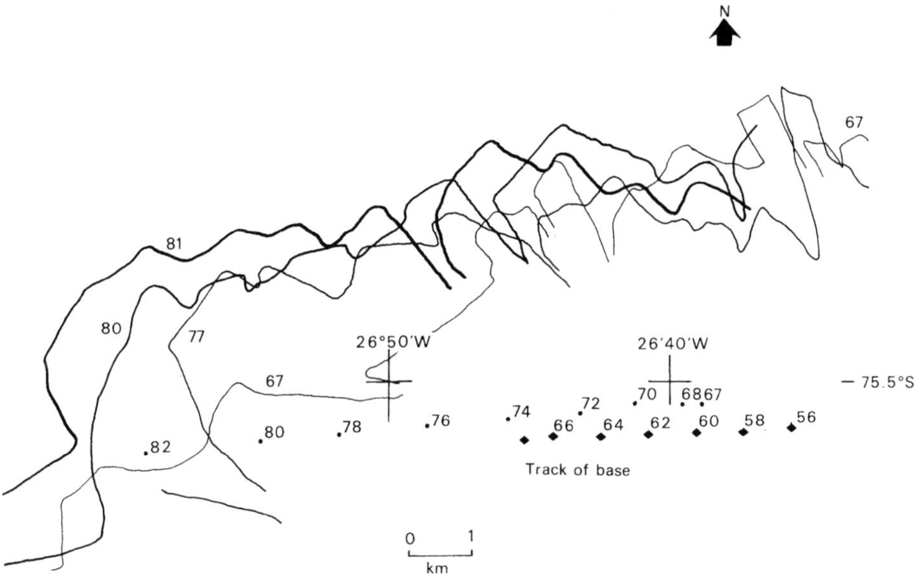

Fig. 2. Position of the ice front in 1967, 1977, and 1981, and successive positions of the station at Halley from 1956–1982.

EDA fluxgate magnetometer (from 1975), with absolute observations being made with a PPM for total field (F), a set of QHMs for horizontal field (H), a PIC declinometer for declination (D) and a BMZ for vertical field (Z). After January 1980 the La Cour variometers, the declinometer and the BMZ were withdrawn, D being measured with the QHM and Z calculated from F and H. All these instruments were mounted on stout wooden posts, fixed in the ice to a depth of 1 or 2 m (Figure 3). Although these instruments can be used in the conventional way, even at emperatures below 0 °C, difficulties arise because:

(i) there is tilting of the wooden posts in the ice, particularly initially; tilting of the ice shelf is also possible,

(ii) the buildings in which the instruments are housed tend to be crushed by the ice; periodically ice has to be chipped away to prevent the roof from coming into contact with the variometers,

(iii) the fact that the building is below surface level requires additional procedures for determining D.

3. Magnetic Topography

Local surveys showed the presence of a region of comparatively large spatial variation in the surface geomagnetic field close to the station in 1960, and again in 1968 after the new station had been built. A grid of stakes at 100-m intervals covering an area

Fig. 3. The fluxgate magnetometer sensor, mounted below the snow surface.

approximately 1 × 2 km was laid out and regularly surveyed, initially with a BMZ and later by PPM. This grid was extended from time to time as the motion of the ice shelf carried it away from the region of magnetic variation. Figure 4 shows this feature in 1974 by which time it was 2 to 3 km east of the station. These surveys, originally intended to provide corrections for the absolute observations, were continued so as to provide information on the movement of the ice shelf.

4. Declination

To find the 'declination' or angle between true North and magnetic North both these directions need determining. Conventionally the instrument used to establish the

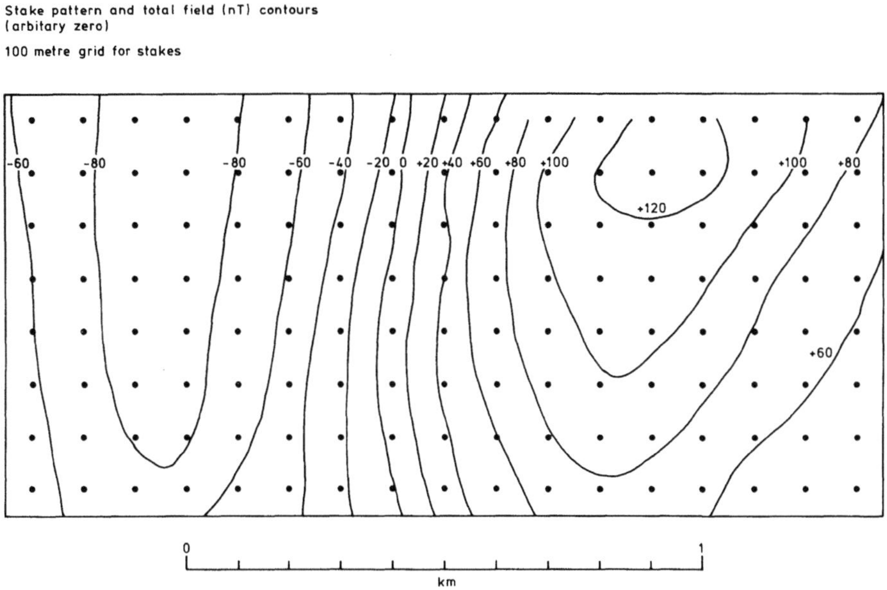

Fig. 4. Magnetic survey in 1974, showing stake pattern and contours of total field.

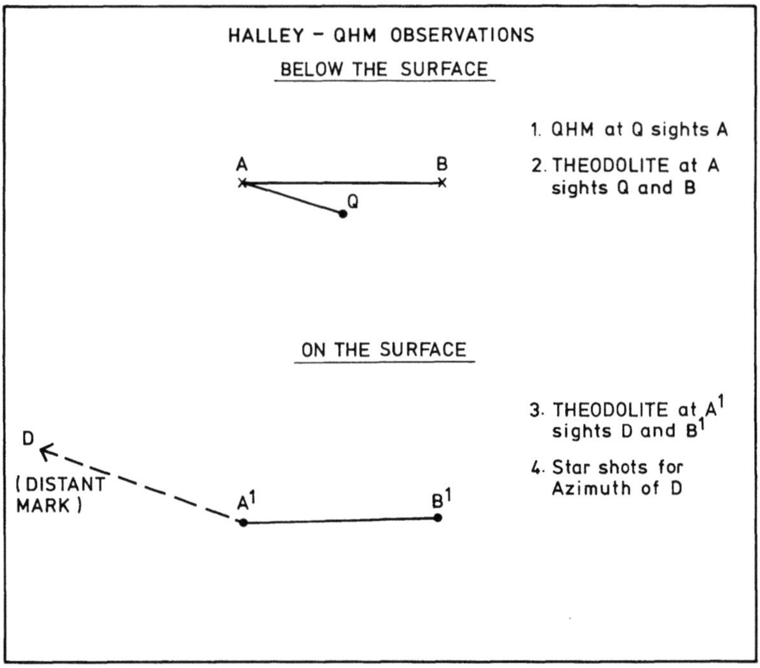

Fig. 5. Diagram showing measurements made above and below the surface at Halley to give declination.

magnetic meridian is also used to sight a fixed mark whose direction (or azimuth), with respect to the Geographic meridian is known and constant.

At Halley, the changing azumuth of a distant mark, also on the ice shelf, is regularly determined from astronomical observations and this information is transferred to the magnetic instrument below the surface; Figure 5 shows the plan above and below the surface. Points *A* and *B* are marks at the bottom of vertical shafts through the ice so that plumb lines may be suspended above them from points *A'* and *B'* on the surface. Two or three times a year a theodolite is used at *A'* to measure the azimuth of the distant mark

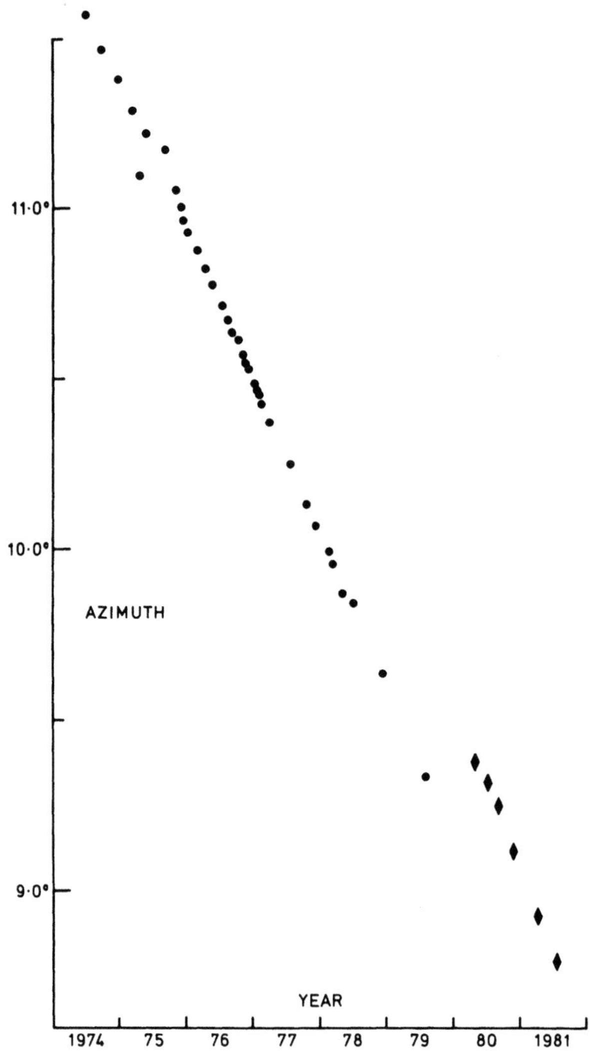

Fig. 6. The azimuth of the distant mark *D* from 1974–1981.

D by means of astronomical observations. The method, taken from Roelofs (1967), uses groups of stars of the same altitude and evenly spaced in azimuth and is relatively unaffected by altitude errors due to atmospheric refraction, etc. It also gives position.

The azimuth of the mark in the period 1974–1982, is shown in Figure 6. The rate of change of azimuth, or rotation, is approximately 30 min of arc yr^{-1}. The discontinuity

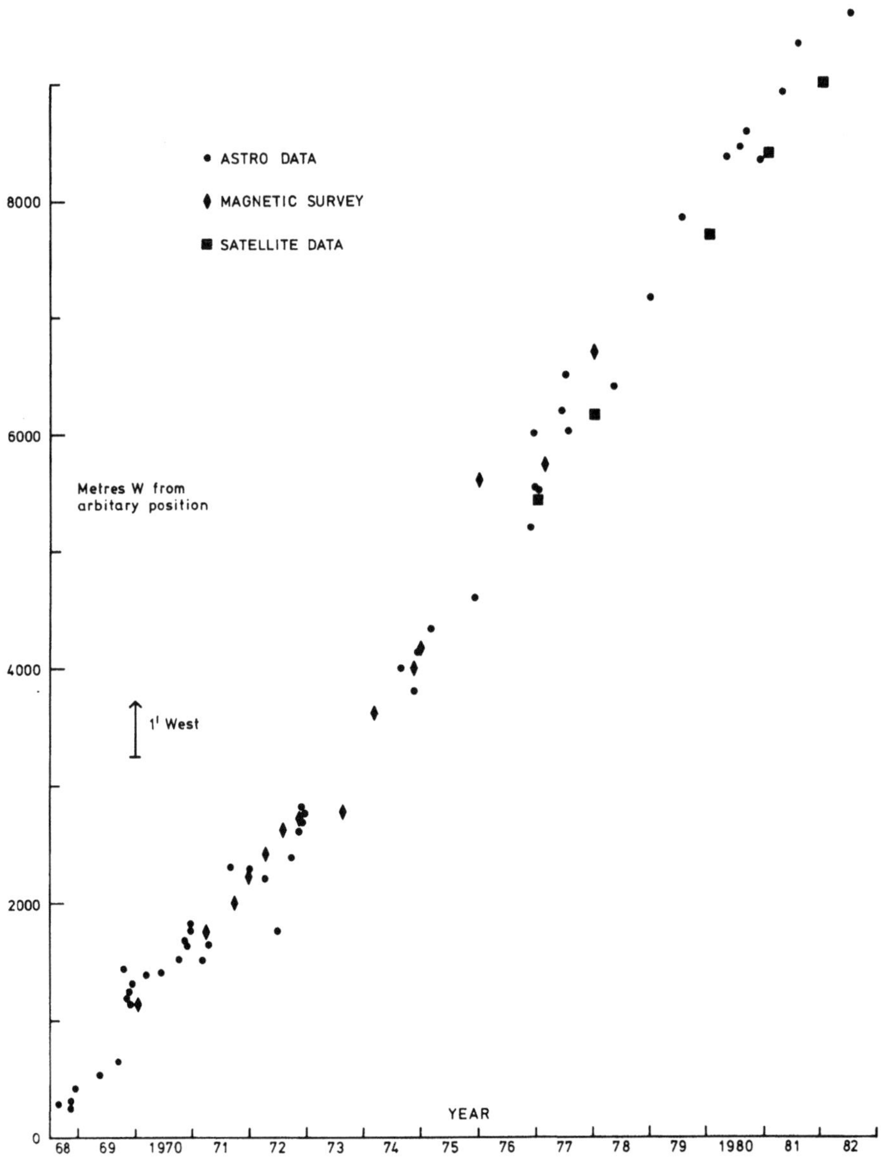

Fig. 7. Position of Halley 1968–1982, in metres West of an arbitrary point.

in 1979/80 is due to the transfer of the instruments to a new location. The QHM could no longer be used in the original 'room' as this was badly crushed by pressure from the ice. It was therefore moved about 100 m to the location previously occupied by the La Cour variometers, and a new shaft dug for BB'.

The theodolite is also used at A' (over A) to measure the angle $DA'B'$, thus establishing the true direction (or azimuth) of the line $A'B'$. Thus a theodolite used below the surface over the point A to measure angle QAB enables the azimuth of QA to be determined. The QHM or declinometer is used conventionally at Q, sighting as an azimuth mark a plumb line over A.

5. Position

Both the magnetic survey and the astronomical observations described above give information on the station position. From 1977 onwards the position of the station was also determined during the annual relief calls by the RRS *Bransfield*. The position of the ship was determined by an onboard satellite navigation system, and the position of a metal mast at the station with respect to the ship determined using the ship's radar.

The results of these position fixes obtained by all methods since 1968 are shown in Figure 7. Figure 8 shows the gradient of a smooth curve drawn through these points using a French curve, from which it can be seen that westward velocity increased from about 400 m yr^{-1} in 1968 to about 800 m yr^{-1} in 1982.

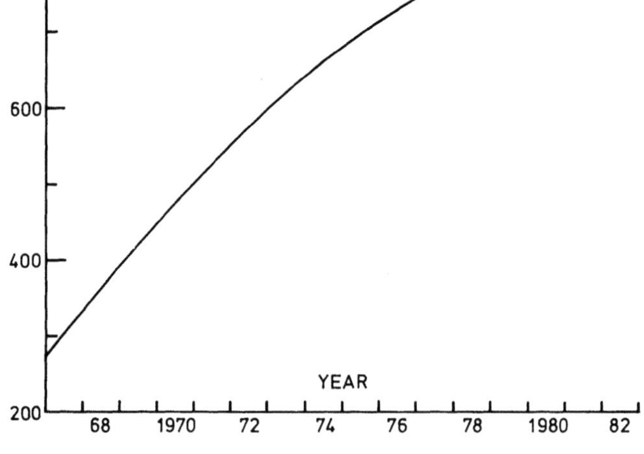

Fig. 8. Rate of westward movement (velocity) 1968–1982.

6. Baseline measurements

Absolute observations were performed at least once a week for most of the period; baselines were determined in the usual way and plotted against time at regular intervals.

The La Cour variometer baselines were well determined, with the scatter of observed values being usually no more than ± 5 nT. The D and Z baselines often showed linear drifts which were steady for periods of up to one or two years. The values given in Table I are for 1977–1979, and are typical of the period 1968–1979.

TABLE I

Baseline drifts at Halley 1977–1979 and 1982

	1977–1979 La Cour	1982 Fluxgate
H	negligible	-86 nT a^{-1}
D	$-29'$ a^{-1}	$-30'$ a^{-1}
Z	-70 nT a^{-1}	$+44$ nT a^{-1}

D is measured positive easterly; Z changes are algebraic.

The fluxgate magnetometer baselines are much more erratic, showing larger scatter, and drifts seldom maintained steady for more than a few months. The values given in Table I are for a few months in 1982.

7. Effects of the Magnetometer Movements on Baseline Values

Since the magnetometer may both rotate and tilt, it is necessary to consider the effects of these movements on magnetometer baselines. These effects have been calculated and are summarised in Table II.

Rotation in the horizontal plane of the ice shelf in which the magnetometers are fixed is equivalent to a change in D for both the La Cour and Fluxgate magnetometer. For example a rotation of one minute of arc westward of the ice shelf will result in a baseline shift of one minute westward, provided that the absolute observations have been corrected for the changing azimuth of the distant mark.

The effect on the baselines of a tilt of the La Cour H and D variometers is negligible, though there may be a deterioration in image quality. Tilt of the Z variometer will give a baseline shift if it occurs about an axis parallel to the axis of rotation of the Z magnet. Since the effect is optical the size of the baseline shift is determined by the distance from the variometer to the recorder drum and by the scale value of the instrument. Tilting the fluxgate magnetometer head alters the direction in which the three measuring axes lie. Thus, if the H sensor is no longer horizontal, it will measure a component of Z in addition to H. The magnitude of the effect thus depends on the relative magnitudes of the H and Z field components.

TABLE II

The effect of magnetometer movements on baseline values.

At Halley $H \simeq 20\,200$ nT, $Z \simeq -41\,500$ nT

Movement		La Cour Scale value $6\,nT\,mm^{-1}$	Fluxgate Scale value $8\,nT\,mm^{-1}$
1′ Tilt about an E-W axis (N Down)	H	—	+12 nT
	D	—	—
	Z	—	+6 nT
1′ Tilt about a N-S axis (E Down)	H	—	—
	D	—	+2′
	Z	+3 nT	—
1′ Rotation (westward rotation of building)	H	—	—
	D	−1′	−1′
	Z	—	—

Note D is measured positive easterly; Z changes are algebraic.

8. Discussion

The La Cour D variometer shows a baseline drift of -29 min yr^{-1}, close to what would be expected due to the rotation of the ice shelf. The La Cour Z baseline drift of -70 nT yr^{-1} could be explained by the variometer tilting 23 min yr^{-1} down to the west. Alternatively, unless the horizontal rotation of the ice shelf is balanced by a similar secular change in D (Brunt, 1960), there will be a steady change in the orientation of the Z variometer magnet with respect to the magnetic meridian. This could cause a change in the Z magnetometer baseline.

For the fluxgate magnetometer, whilst the D baseline drift has usually been close to that caused by rotation, drifts for H and Z have been more variable. In particular, to explain the drifts shown in Table I, a 7 min tilt down to the S is required for the H variometer, and a 7 min tilt down to the N for the Z variometer. Other possible explanations are temperature or other drifts in the magnetometer electronics and the effect of moving over an area where there is a large spatial variation in the magnetic field. Changes of the field that are equal at the sites of both the absolute and the recording instruments should not effect the baselines. For such field changes to cause baseline drifts the difference in field between the absolute and recording instruments has to change. At Halley these instruments are separated by no more than 10 m, so this does not seem to be a likely cause of the observed baseline drifts.

The difficulty in explaining properly all the baseline drifts and the fact that the observatory has been moving in a region of appreciable magnetic gradient makes the derivation of reliable absolute values very doubtful. It is therefore our intention to treat Halley as a geomagnetic variation station.

Acknowledgments

We wish to thank the numerous staff in the Antarctic who have made the observations, sometimes in extreme weather conditions, and our colleagues in Cambridge for suggestions, criticism and help with the necessary computing.

References

Brunt, D.: 1960, *The Royal Society for International Geophysical Year Antarctic Expedition Halley Bay, Coats Land, Falkland Islands Dependencies, 1955–1959*, Vol. 1, Sir David Brunt (ed.), Royal Society, London. pp. 61–124.

Roelofs, R.: 1967, *In International Dictionary of Geophysics*, S. K. Runcorn (ed.), Pergammon Press, Oxford. pp. 588–589.

The last group of papers addresses one of the consequences of digital recording at geomagnetic observatories – the production of indices. On one hand the judgement of an experienced observer using analogue records can produce an index free from some of the quasi regular variations. This takes time and indices prepared in this way are often two or more weeks in arrears. Digital indices can be produced at the same time as the field record, they can be stored on file at once and made available to users via computer networks. Using advanced technology both records and indices can be available in real time. Computers cannot make judgements, so computer indices will differ from traditional ones. These papers investigate the range of errors of the traditional K index and relate differences between a likely computer generated equivalent K index to the error distribution in K itself. The scientific community must now decide which it prefers. This set of papers also prompts the scientific community to consider whether the service of indices it receives from Geomagnetism is sufficient for its needs or whether, in view of the ease of index generation by computer, new indices should be defined.

PROPOSED MAJOR FORMAT CHANGE TO GEOMAGNETIC ACTIVITY REPORTS AND FORECASTS PRODUCED BY THE SESC, BOULDER, COLORADO, U.S.A.

J. A. JOSELYN

Space Environment Services Center, NOAA R/E/SE2, 325 Broadway, Boulder, CO 80303 U.S.A.

Abstract. The Space Environment Services Center (SESC) provides real-time information on the state of the solar-terrestrial environment. In particular, the geomagnetic field is closely monitored. Summaries of geomagnetic activity are issued daily, as are forecasts for each of the next three days. Customers are alerted when activity rises above selected thresholds of the *K*- and *A*-indices. This present geomagnetic reporting and forecasting format is no longer the most appropriate for the expanding number of agencies and industries which the SESC serves. These customers need more detailed information about the geographic distribution of geomagnetic activity stated in plainer language. The proposed new format would quantitatively describe geomagnetic activity in terms of 6 qualitative categories: quiet, unsettled, active, minor storm, major storm, and severe storm. Summaries and forecasts would be made for 2 specific latitude zones: mid-latitudes (20–50 deg geomagnetic) and high-latitudes (50–80 deg geomagnetic). The format adapts well to longer-range (27-day) forecasts.

The Space Environment Services Center (SESC), a daily 24-hr joint operation of the National Oceanic and Atmospheric Administration (NOAA) and the US Air Force Air Weather Service, collects data and provides alerts and forecasts on solar activity and the effects of that activity on spacecraft and a variety of terrestrial systems. In particular, the geomagnetic field responds to variations in the solar wind and severe disturbances (geomagnetic storms) can result. Auroras are a familiar result of geomagnetic activity, but the same solar-terrestrial interaction which drives the auroras induces currents in power lines, pipelines and cables, and can compromise geomagnetic survey expeditions and high-frequency radio communications and radars. Joule heating increases upper atmospheric temperatures and densities leading to increased satellite drag at low altitudes. At geosynchronous altitudes, a phenomenon known as spacecraft charging can totally disable satellites. Finally, it has been shown that homing pigeons and other migratory creatures rely on the geomagnetic field as part of their intrinsic navigation systems, and they can become disoriented and permanently lost during storms.

The SESC has been likened to a 'space weather' agency. Observations of solar activity, the interplanetary medium, the magnetosphere, ionosphere, and the geomagnetic field at widely-spaced locations are continuously fed into a centralized computer system, where they are stored, analyzed, and interpreted. Every day at 2200 UT, the SESC issues a "Joint USAF/NOAA Primary Report of Solar and Geophysical Activity" which summarizes the day's events and predicts the possibilities for major solar flares and the level of geomagnetic activity for each of the next three days. Table I briefly lists the variety of data which composes our unique data base. Figure 1

Geophysical Surveys **6** (1984) 419-424. 0046-5763/84/0064-0419$00.90.

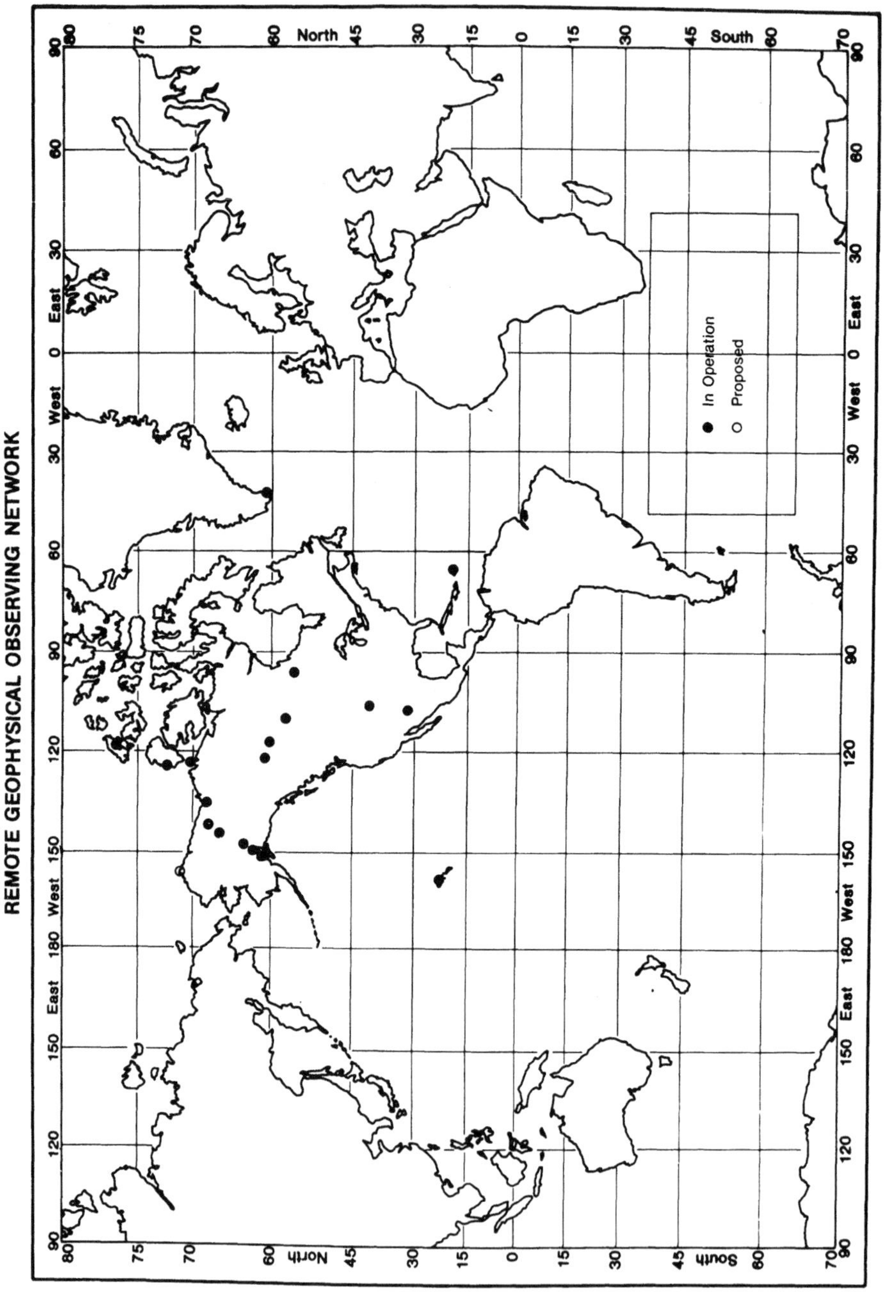

Fig. 1. Geomagnetic observatories which contribute data in near real-time.

TABLE I

Abbreviated list of SESC Solar-Geophysical Data

Solar Observations
continuous X-ray flux measured at two geosynchronous satellites
continuous optical and radio noise patrol using a worldwide network
daily sunspot drawings, magnetograms, and Hydrogen-alpha and other spectroheliograms

Energetic Particle Measurements
continuous fluxes of high-energy protons (> 0.8MeV) and electrons (> 2 MeV) measured at two geosynchronous satellites
continuous data from a neutron monitor and a riometer at Thule, Greenland, and other riometers in Alaska

Magnetometer Data
continuous vector data from 17 ground-based observatories
continuous vector data from 2 geosynchronous satellites
15-min summary data from 4 other ground-based observatories
daily summaries from the worldwide IUWDS magnetometer network

Ionospheric Data
hourly total electron content data from 11 sites
hourly high-frequency propagation parameters (e.g. foF2) from 8 sites
daily summaries of ionospheric parameters from the worldwide IUWDS network

Interplanetary Data
intermittent plasma and magnetic field data from the International Cometary Explorer (formerly ISEE-3)

specifically shows the location of the magnetometers which contribute vector data in near-real-time to the SESC for use in diagnosing the state of the geomagnetic field. Magnetospheric magnetometer data are also telemetered from 2 of the NOAA/GOES geosynchronous satellites. The SESC distributes data, indices, summaries and forecasts in a number of ways. These include a continuous direct satellite broadcast, teletype, telephone (including on-duty forecasters or technicians and a recorded message), a summary message updated every 3 hr and broadcast over WWV radio frequencies, and publications sent by mail. It is also possible to interact directly with the data base computer.

Since 1966, the Space Environment Services Center (SESC) has reported and forecast A-indices as an indicator of geomagnetic activity. The A-index is a daily average of the eight, 3-hr a-indices which approximately represent half of the gamma range of the most disturbed field component at a mid-latitude station. In practice, the a-index is obtained by simple conversion from the measured 3-hr K-index, which ranks geomagnetic activity during the Greenwich day according to increasing disturbance using a scale from 0 (quiet) to 9 (intense storm). In SESC products, two A-indices are reported and forecast: A Fredricksburg (i.e. the A-index measured at Fredricksburg, Virginia) and an estimated Ap, a real-time index which closely approximates the IAGA geomagnetic planetary index determined by the Institut für Geophysik located at

Göttingen, Germany. The A-index is well-defined but best serves a scientifically-sophisticated user and is no longer appropriate for the expanding number of agencies and industries which the SESC serves. In addition, present forecasts are do not imply the level of confidence in the forecast, and they do not address the well-known distribution of geomagnetic activity with latitude.

At the 1979 Solar-Terrestrial Predictions conference, it was strongly recommended that high-latitude geomagnetic activity be reported and forecast separately from those for mid-latitudes (Working Group Report on Geomagnetic Applications, 1979). The rationale for separation can be seen by inspection of Figure 2, redrawn from Campbell (1973), which shows the average amplitude variation of the four solstitial and equinocial months for 1965 for the three magnetic field components and the total field. At least four zones of activity can be identified: equatorial (00–20 deg geomagnetic), mid-latitude (20–50 deg), high-latitude (50–80 deg) and polar (80–90 deg geomagnetic). Because most of the SESC customers reside between 20 and 80 deg, it was decided to forecast only for those two zones at this time. The high-latitude index would be specifically useful for transpolar radio communication and radar operations, Alaskan and Canadian geophysical surveyors, auroral zone research campaigns, and possibly for satellite drag and other studies of the neutral atmosphere. An original candidate for the high-latitude index was the AE (Auroral Electrojet) index, first

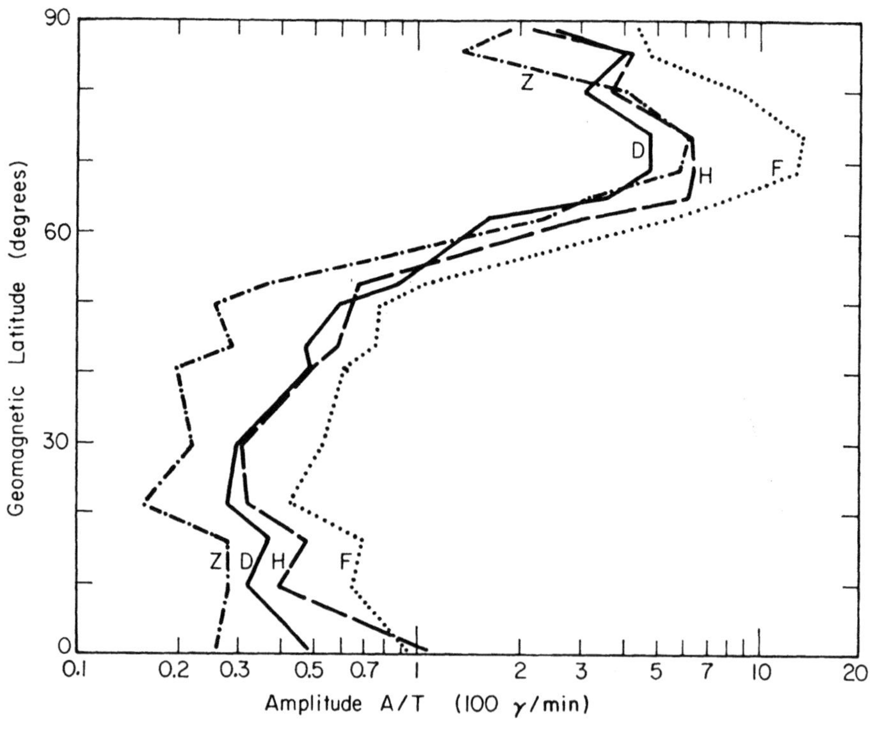

Fig. 2. Magnetic field components (H, D, Z) and total field (F) from Campbell (1973).

derived by Davis and Sugiura (1966). However, the standard AE is an 'instantaneous' global index constructed from 11 or 12 observatories approximately equally spaced around the auroral oval. Among other difficulties with the *AE* index, it was decided that the SESC magnetometer network does not cover enough longitude to adequately estimate or predict this number. At this time, the most practical high-latitude indices available to the SESC are the observed *K*- and *A*-indices from College, Alaska. The reports and forecasts for this station could be handled in an analogous way to those for Fredricksburg, Virginia (our mid-latitude 'standard' station). College data is received in near-real-time in Boulder, and *K*-indices have been scaled continuously since 1941, providing a good climatological data base. Unfortunately, *AE* records show that College is often south of the most intense auroral activity, indicating that we should pursue acquiring data from a more northerly site in the future. Other possibilities for a high-latitude index are being explored by the Canadian Department of Energy, Mines and Resources in the Division of Seismology and Geomagnetism. They are developing a detailed report and forecast of high-latitude geomagnetic activity, to include 3 high-latitude zones: subauroral (50 to 60 deg geomagnetic), auroral (60 to 70 deg), and polar (70–90 deg) (E. R. Niblett, 1983, private communication).

As a replacement to the present format, we propose to provide probability forecasts of expected geomagnetic conditions by category for two geomagnetic zones: mid-latitudes (20 to 50 deg geomagnetic, as represented by Fredricksburg) and high-latitudes (50–80 deg geomagnetic, as represented by College, Alaska). An example is shown in Table II. To arrive at this output forecast, the forecaster will actually begin with a much more detailed computer-assisted worksheet. Climatology (the average conditions expected solely on the basis of historical data) and persistence forecasts will be augmented by forecaster skill to produce the final forecast. Numerical models of events (e.g. major flares) can be easily incorporated. Probability forecasts will be verified against observed conditions at the representative stations using the Ranked Probability Score method.

We have chosen a probability format for several reasons. First, customers need not understand the *A*-index *perse* to use the forecast, since the names of the categories are indicative of geomagnetic conditions. Second, probability forecasts reveal forecaster confidence. In Table II, if the probabilities were distributed more evenly the forecast would clearly be less certain. Third, customers can have widely-varying priorities. With

TABLE II

Sample probability forecast of geomagnetic conditions

Category	Quiet	Un-settled	Active	Minor storm	Major storm	Severe storm
A Ranges	0 – 7	7 – 15	15 – 30	30 – 50	50 – 100	100 – 400
Mid-latitude	10	80	10	0	0	0
High-latitude	10	70	20	0	0	0

probabilities, they can construct individual cost-loss decision matrices, and act only when the forecast exceeds their critical probability thresholds. Fourth, dramatically changing conditions expected in the middle of a given day can be indicated with a bimodel distribution of probabilities. Finally, probability geomagnetic forecasts would be consistent with SESC flare forecasts and with NOAA National Weather Service products.

A beginning date for issuing the new format has not yet been chosen. Development is still underway, and a period of forecaster training and familiarization will be necessary. However, it is expected that the switch will occur during 1985.

References

Campbell, W. H.: 1973, 'Spectral Composition of Geomagnetic Field Variations in the Period Range of 5 min to 2 hr as Observed at the Earth's Surface', *Radio Science* **8**, 929–932.

Davis, T. N. and Sugiura, M.: 1966, 'Auroral Electrojet Activity Index *AE* and Its Universal Time Variations', *J. Geophys. Res.* **71**, 785–801.

Working Group Report on Geomagnetic Applications, W. H. Campbell, Chairman: 1979, *Solar-Terrestrial Predictions Proceedings*, Vol. 2, R. F. Donnelly (ed.), U.S. Dept, of Commerce, NOAA, pp. 133–136.

ABOUT DIGITAL ALTERNATIVE TO THE *Kp*-INDEX

ASGER LUNDBAK
*Copenhagen**

Abstract. For more than 40 yr the so-called K-indices have been used to characterize irregular geomagnetic activity. Now that new methods of recording and digital computers are at our disposal it may be questioned, whether some alternative procedure for characterizing of irregular geomagnetic activity has to be preferred.

A proposal for a new procedure is presented. The indices such obtained are named aK (meaning 'alternative K'), and some conformity with the old K-indices is aimed at. It is not attempted to distinguish between wave radiation and corpuscular radiation from the Sun; consequently, the so-called S_R variation is not removed.

Elaboration of the procedure is realized by inserting a few extra statements in an extensive algorithm for general processing of digitally recorded data. A gradual modification of these statements, while working with the computer language PL/1, has lead to the final formulas.

An appendix is added at the end of the paper. It is a condensed version of the procedure together with a – possibly – clarifying example.

1. Introduction

In 1939 the so-called K-index was adopted as a measure of irregular variations in the geomagnetic field (Bartels, 1940, pp. 339–343). The rules for derivation were such, that the K-values obtained might represent the corpuscular radiation from the Sun, and as far as possible be independent of the solar wave radiation.

The K-index is expressed in integers from 0 to 9, and it may be considered as a code rather than as a real measure. Further, the Kp-index is a standardized mean of the K's from a selected group of observatories. These Kp's are truncated to integers as well as to intermediate values containing one and two thirds of a unit, normally presented in this way: 0o, 0+, 1−, 1o, 1+, 2−, 2o etc.

Although the Kp-indices have played a very basic role in geomagnetism and aeronomy during almost half a century, it may now be questioned, whether these indices really fulfil their original purpose in a fully objective way. Furthermore, the methods for recording of geomagnetic components are being changed at many observatories, from photographic to digital recording. Automatic systems are in use, which provide digital values on magnetic tape. Therefore – thanks to modern computers – procedures to derive appropriate indices can now be less time-consuming and possibly more exact than hitherto.

A few drawbacks of the traditional procedure used for the derivation of K-values may be mentioned. One drawback is, that a choice has to be made between D- and H-components instead of using their combined amplitude (cf. Figure 1). Another drawback is the kind of scale, which is used; its steplike nature may at times result in

* Now retired; private address: Stigaardsvej 3, DK-2900 Hellerup.

Geophysical Surveys **6** (1984) 425-430. 0046-5763/84/0064-0425$00.90.

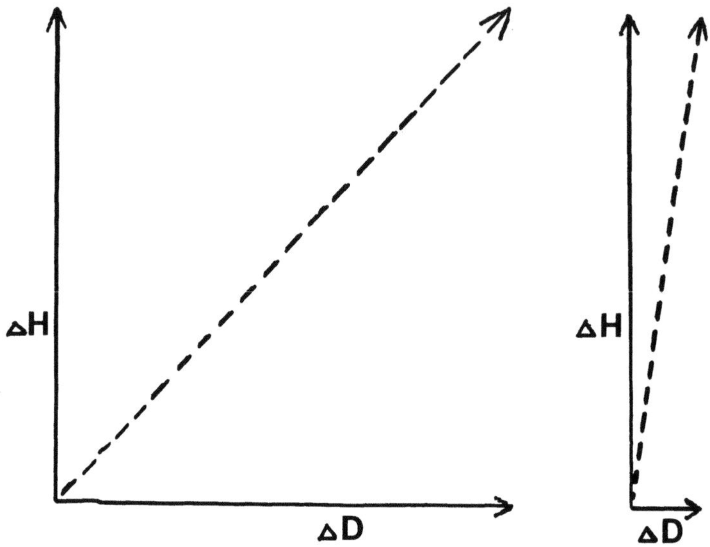

Fig. 1. It is a weak point in the traditional K-procedure, that only one of the horizontal components is used in each case, namely that component, which has varied most during the 3-hr interval in question. Here, the H-component has to be used in both cases represented; but it appears to be a better choice to use the combined variability of both components, i.e. the stippled resultants.

inconsistent values. Also, it is a serious objection, that subjective evaluation can play a role.

It appears evident – especially because of the subjectivity, which is incorporated in the manual procedure – that some other indices without the drawbacks mentioned ought to be defined. It is possible to adjust such new indices, so that there will be no significant discontinuity, when use of the new indices is adopted.

The present author has tried to produce new indices, applying a few relevant formulas. No presupposition about corpuscular and wave radiation is made in the derivation of these alternative indices. Their only purpose is to be an objective measure of the irregular geomagnetic activity, because a clear distinction between corpuscular and wave radiation may often be rather doubtful. The symbol aK is used for the indices chosen.

The procedure for obtaining of the aK-values contains two steps or phases. In the first phase, which may be named the definition phase, a process is stated for extraction of figures representing the geomagnetic activity hour by hour. The second phase could be named the conversion phase, the aim being to transform the figures mentioned to fit into a scale like the Kp-scale. This second phase is purely empirical.

2. Procedure

Hourly ranges r_H and r_D (or r_X and r_Y) of the horizontal components are used in the first phase, and nanotesla (nT) are supposed to be the units. The values of the ranges are slightly modified, namely by subtracting 5 nT, if the value in question is greater than 10 nT, and by halving the value, if it is 10 nT or less. This kind of modification, which may appear physically unfounded, is a first step to bring the scale for the *aK*'s near to the *Kp*-scale. The modified ranges are named *Hr* and *Dr*.

Dr and *Hr* have to be inserted in Equation (1), and here a modification more is undertaken, namely by applying the factor $600/U$. U is a quantity depending on the geomagnetic position of the observatory in question, and the so-called 'lower limit' value (Mayaud, 1980, p. 42) may be used. For instance, this value is for Niemegk 500 nT, for Rude Skov (now replaced by Brorfelde) 600 nT, and for Lerwich 920 nT. Alternative values – preferred by the author and deviating slightly from the above values – can be derived by the formula $U = 10\,000/(70 - v)$, v being the numeric value of the corrected geomagnetic latitude (Mayaud, 1980, pp. 137–138). Using for Niemegk, Rude Skov and Lerwich the geomagnetic latitudes 49.2°, 53.3°, and 59.4° the U-values will be 481, 599, and 943 nT, respectively.

$$R = \frac{600}{U} \times \sqrt[2]{(Dr)^2 + (Hr)^2} \tag{1}$$

$$S = \left(\frac{R}{a \times \sqrt[R+1]{b}}\right)^c \tag{2}$$

$$aK = S \times \left(d - \sqrt[e]{|S - 2| + 0.1}\right). \tag{3}$$

Equation (2) and (3) deal with the second phase of the procedure. The highest of the R-values from 3 consecutive hours is chosen as a representative value for the irregular geomagnetic activity in the whole interval of 3 hr. Now, the aim of the two formulas is to convert the values such chosen to new values, which come as near as possible to K-values derived by the old procedure.

This conversion was made by introduction of an extra segment in a rather extensive algorithm for general processing of geomagnetic data. A gradual adjustment of the segment took place, and the final statements in the segments can be expressed mathematically in Equations (2) and (3).

The constant terms in the formulas have got their final values by a 'trial and error' method (as to the term 0.1 in Equation (3) its function is to prevent so-called underflow as well as overflow during computing). As to the constants, which are represented by letters in the formulas, the following values are arrived at: $a = 7$, $b = 10^8$, $c = 0.4$, $d = 3$, $e = 3$.

A concrete version of the procedure is – as mentioned earlier – given in Appendix A together with an example, which can be controlled by means of a simple pocket computer. Insertion in Equation (1), (2), and (3) of the above constants is made in the appendix.

3. Conclusions

Comparisons have been made using photographic recordings from the Danish Rude Skov observatory to produce traditional K-values, and using digital values from the new Danish observatory, called Brorfelde, to produce aK-values. The two types of indices have been derived quite independently of one another. The acronyms and the geographic coordinates of the observatories are:

RSV	55° 50.6′ N	12° 27.4′ E
BFE	55° 37.5′ N	11° 40.3′ E.

In the first instance, differences have been computed between three kinds of daily totals: aK's determined by the formulas, K's determined in the traditional way, and Kp's published by IAGA. It must be remembered, that the principles, on which the aK-computing is based, deviate somewhat from the K- and Kp-principles, so that full agreement cannot be expected; but the same level is aimed at. However, according to the author's experience the level of the aK's is slightly higher than the level of the Kp's during some months. Possibly, a rounding to the largest integers, which are less than or equal to the original aK's, may be reasonable, when tenths or still smaller parts of indices are not wanted.

On the assumption, that the daily Kp-totals are considered as 'true' totals, the standard deviations from these values of the aK- and K-totals have been computed for each of 5 months, which were picked out from different seasons of the years in question. The results are stated in the following scheme.

Months and years	Daily aK-totals versus Kp-totals	Daily K-totals versus Kp-totals
September 1979	2.75	3.07
May 1980	2.81	2.01
February 1982	2.18	7.62
May 1982	2.69	4.00
July 1982	3.66	4.16

Further, the reliability of the aK's has been evaluated by using single values, i.e. indices not added together. For each month one and the same interval of 3 hr is chosen for comparison of the aK-, K-, and Kp-indices. Again, the aK- and the K-values are subtracted from the Kp-values, and the differences are analyzed. As tenths of integers are used in the aK's, integers exclusively in the K's, and thirds of integers in the Kp's, all of the more than 200 differences of each kind have been rounded to their nearest integers. A summing up of these differences from 5 months is presented in Figure 2.

In conclusion it may be stressed that the alternative index here presented does not pretend to be a copy of the K- or Kp-index. In the broad outline this new index is in accordance with the Kp-index; but the single values of aK and Kp may to some extent deviate from one another for various reasons.

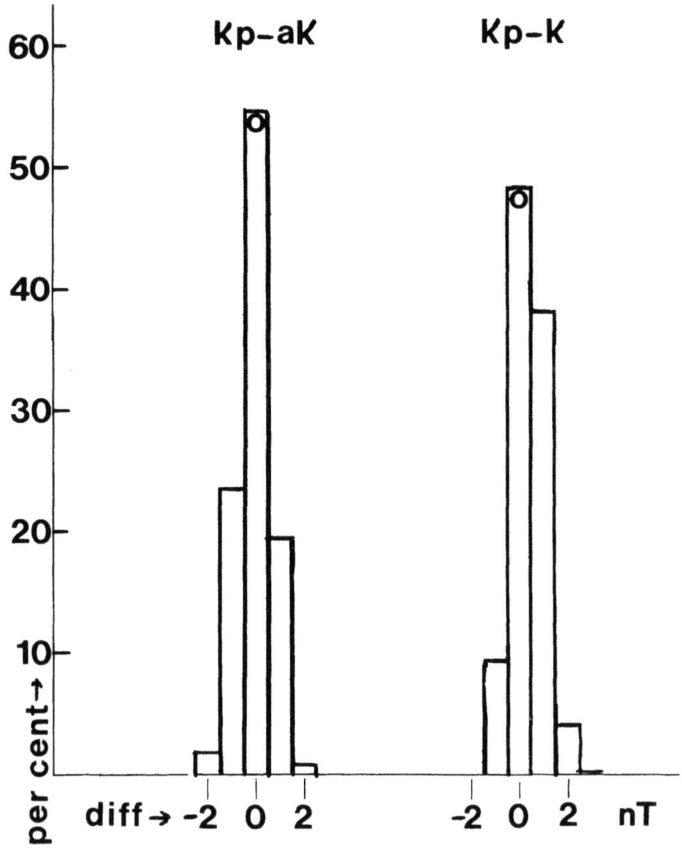

Fig. 2. The results obtained from the 5 months comparison are illustrated. Rounded differences between the official Kp-values and the computed aK-values are represented to the left, whereas similar differences between Kp-values and the manually produced K-values are represented to the right. In both cases more than 200 differences have been used. Considering the frequency in per cent of the various differences it appears, that the agreement is a little better in the first case than in the second one.

Appendix A: Computation of aK

(*Extract of IAGA-paper, August* 23, 1983)

The formulas below have been chosen to compute an alternative index aK to replace the traditional K-index. This kind of alternative index may be direct output from a computer, when the geomagnetic components are recorded digitally.

In the first instance a special value for the observatory in question (corresponding to the lower-limit value for $K = 9$) has to be determined by the formula $U = 10\,000/(70 - v)$, v being corrected geomagnetic latitude. Further, the hourly ranges in nT of D and H (or Y and X) are reduced by 5 nT, if they are greater than 10; if not, they are divided by 2. These modified ranges – named *Dr* and *Hr* – as well as *U* are inserted in the simple formula:

$$R = \frac{600}{U}\sqrt{(Dr)^2 + (Hr)^2}.$$

Further, for each 3 hr the greatest of the *R*'s is used in the following two formulas:

$$S = \left(\frac{R}{7\sqrt[R+1]{10^8}}\right)^{0.4}$$

$$aK = S\,(3 - \sqrt[3]{\text{abs}(S-2) + 0.1}).$$

Example: $v = 53.3°$, $U = 599\,\text{nT}$

	Interval 12–15 UT			Interval 15–18 UT		
D-max.	340	327	333	349	476	403
D-min.	319	313	315	330	323	226
Dr =	16	09	13	14	148	172
H-max.	17219	17242	17264	17270	17718	17627
H-min.	17195	17193	17189	17192	17215	17302
Hr =	19	44	70	73	498	320
R =	24.9	44.9	71.3	74.5	520.4	363.9
S =			2.285		5.525	
aK =			5.2		8.1	

Comment: In fact, the applied figures are from the Danish Brorfelde observatory on July 13, 1982. For the two intervals the corresponding *K*-values (measured in traditional way) are 4 and 7, whereas the *Kp*-values (published by IAGA) are 4+ and 8+, respectively.

References

Bartels, J.: 1940, 'Solar Radiation and Geomagnetism', *Terr. Magn. Atmos. Elec.* **45**.

Mayaud, P. N.: 1980, *Derivation, Meaning and Use of Geomagnetic Indices*, American Geophysical Union, Washington D.C.

DERIVATION OF *K*-INDICES USING MAGNETOGRAMS CONSTRUCTED FROM DIGITAL DATA

E. R. NIBLETT, E. I. LOOMER*, R. L. COLES, and G. JANSEN VAN BEEK
Division of Seismology and Geomagnetism, Earth Physics Branch, 1 Observatory Crescent, Ottawa, Canada K1A 0Y3

Abstract. The purpose of this study is to investigate the feasibility of deriving the traditional *K* index from magnetograms plotted from recorded digital data. Digital magnetic data from Ottawa Observatory are available for the *X*, *Y*, and *Z* components in the form of 1 min averaged values and spot values at selected sampling intervals of 1 sec or greater. It is found that the lower *K* values tend to be biased downward by one level when a digitizing interval greater than 30 sec is used for construction of the artificial magnetograms. However, for digitizing intervals of 30 sec or less the reconstructed analogue plots are just as reliable as standard-run Ruska magnetograms for the derivation of *K*.

1. Introduction

The three hour range index *K* was adopted internationally in 1939 and has found wide application in studies of geomagnetism. It has proven particularly useful in investigations of magnetic effects associated with solar wind and the physics of the Sun-Earth environment. The *K* index was developed for use with standard run, normal sensitivity photographic magnetograms produced on a daily basis. Analogue systems of this kind, with a standard recording speed of 20 mm hr^{-1} have been the traditional instrumentation in all countries contributing *K* index data to IAGA for the derivation of *Kp* and other parameters. Over the past decade, however, digital data acquisition has become commonplace and is now supplementing or even replacing the old standard photographic systems at many magnetic observatories. In Canada, for example, Ruska photographic variometers have continued in service at only four observatories in the 12 station network, though *K* indices are still scaled from analogue magnetograms in the time-honoured fashion.

For a variety of reasons it is becoming increasingly difficult to maintain the old analogue systems and these will probably be phased out at all Canadian observatories during the next few years. It is therefore important to examine the precision with which K indices may be derived from magnetograms constructed from digital data and to determine the manner in which this precision is influenced by recording parameters such as the digitizing interval and the filtering used for smoothing and anti-aliasing.

To implement this study *K* values derived from Ottawa standard Ruska magnetograms for 1982 have been compared with plots of averaged 1 min data (AMOPLOTS) derived from the AMOS Mark III digital system (Trigg and Nandi, 1984) operating at

* 26 Willeth Avenue, Nepean, Ontario, Canada K2E 5C3
Contribution from the Earth Physics Branch No. 1132.

Geophysical Surveys **6** (1984) 431-437. 0046-5763/84/0064-0431$01.05.

Ottawa Observatory. The Ruska system recorded H, D, Z components, while the AMOS system recorded X, Y, Z. Calibration constants for both systems were very carefully checked throughout the 12 month period. All digital data were plotted on a scale of 5 nT mm^{-1} with a 'recording speed' of 20 mm hr^{-1}. No attempt has been made to investigate the generation of K indices by computer methods. The sole purpose of this study is to compare K values derived in the traditional way from photographic magnetograms with those similarly derived from analogue plots of digitally recorded data.

2. Data Sampling and Analysis

In the AMOS Mark III system the A–D converter takes three rapid samples each second. The median of these three values is chosen to represent the data for that second. A set of successive median values makes up the 1 sec fast sampling data. This is the basic raw data set produced by AMOS Mark III. The system response of the magnetometer is flat over the frequency range from D.C. to 2 Hz.

In the AMOS software package two filtering algorithms are applied to the fast-sample data to produce a 1 min data set. These provide smoothing for noisy or spiky raw data and eliminate or reduce aliasing problems arising from natural signals at periods $\leqq$ 120 sec.

The first algorithm, which operates directly on the 1 sec fast-sample data, is defined by

$$y_t = y_{t-1} + \frac{x_t - y_{t-1}}{C} \tag{1}$$

where

y_t, y_{t-1} are the outputs at times t, $t-1$
x_t is the instantaneous sample at time t
C is a constant $= 10$ in the AMOS system.

The second algorithm is simply the arithmetic mean of seven successive values outputted from Equation (1) at 10 sec intervals, and centred on the minute, i.e.,

$$z_t = \tfrac{1}{7}\,\{y_{t-30} + y_{t-20} + y_{t-10} + y_t + y_{t+10} + y_{t+20} + y_{t+30}\}. \tag{2}$$

The attenuation curves for each of these filter algorithms, and their combined response, are shown in Figure 1 (Coles, 1983). The effect of the first-order recursive filter (algorithm 1) is not severe at periods greater than 100 sec. However the box-car filter which produces the 1 min averages attenuates short period data very strongly. Little can be done to avoid this problem as long as 1 min values are required for storage and archiving facilities in world data centres. Unfiltered or unaveraged 1 min spot values would still lead to severe reduction of short period fluctuations and would be unsuitable for time-series applications and spectral analysis because of aliasing problems.

Figure 2 shows a standard Ruska H trace recorded at Ottawa on September 5, 1982. A sudden commencement occurred just before 2300 UT. Compared with this photographic magnetogram are six analogue X component plots constructed from AMOS

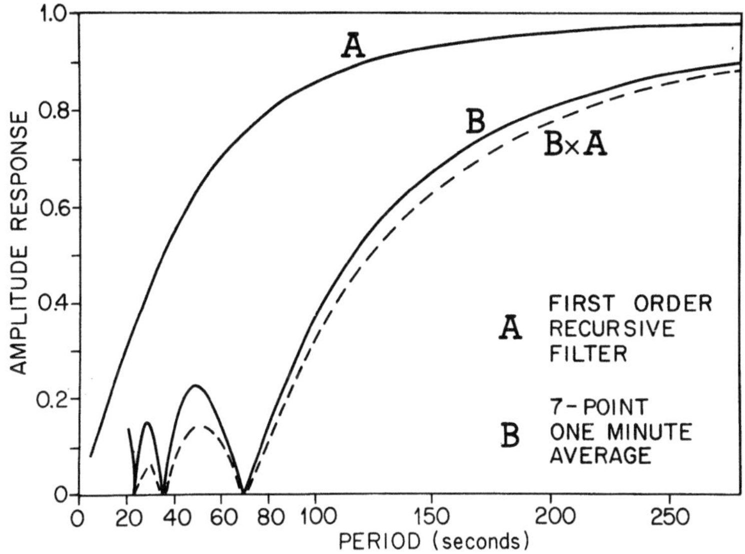

Fig. 1. Amplitude response of the two filter algorithms used in the AMOS III software to produce 1 min averaged values of the recorded geomagnetic components (X, Y, Z, F).

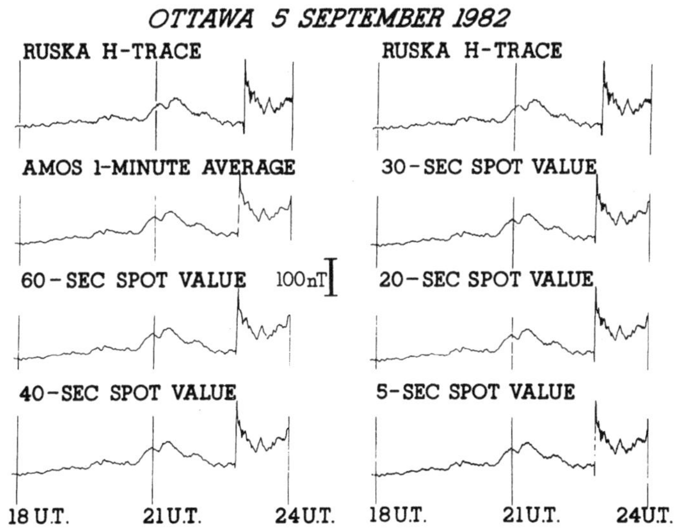

Fig. 2. Comparison between a standard Ruska photographic H component record and analogue plots of AMOS III X component data for various digitizing intervals.

III data. The plot of 1 min averaged data is the AMOPLOT referred to above. The 1 min averages were derived from the 1 sec raw data set after application of algorithms (1) and (2). The remaining plots are unaveraged and unfiltered spot values for the sampling intervals indicated. At Ottawa the declination is approximately 14 °W so that the X and

H traces closely resemble one another. It is readily seen that the 5 sec spot value plot most accurately reproduces the photographically recorded variations including the SC event. On the other hand both the 1 min averages and the 60 sec spot values severely attenuate higher frequency events which show plainly on the Ruska trace.

Standard sensitivity Ruska magnetograms in H and D from Ottawa for 1982 were carefully scaled for K as were the reconstructed X and Y magnetograms plotted from the 1 min averaged AMOS data (AMOPLOTS). The regular magnetic variation S_R was removed from both data sets using normal procedures (Mayaud, 1965a, b). We have denoted K indices derived from Ruska standard magnetograms K_R and those from AMOPLOTS K_A. All intervals for which K_R and K_A were not equal were carefully rechecked and rescaled. In the final analysis K_R was found to exceed K_A in just over 10% of the 2685 intervals measured. K_A was greater than K_R in three intervals only. The difference between K_R and K_A was never greater than one.

The diurnal distribution of discrepant K values (Figure 3) shows that a maximum number occurs at 18–21 UT, with a secondary peak at 9–12 UT. The seasonal

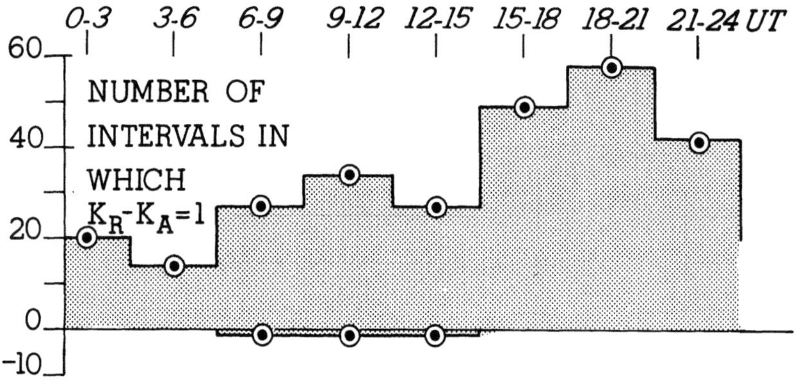

Fig. 3. Diurnal distribution of intervals for which $K_A \neq K_R$ at Ottawa in 1982.

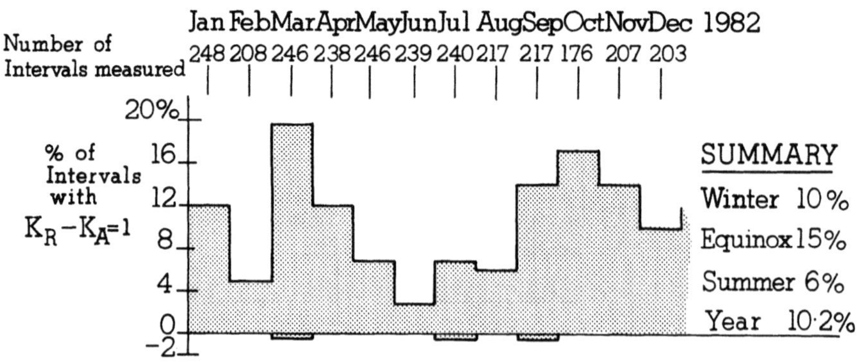

Fig. 4. Seasonal distribution of intervals for which $K_A \neq K_R$ at Ottawa in 1982.

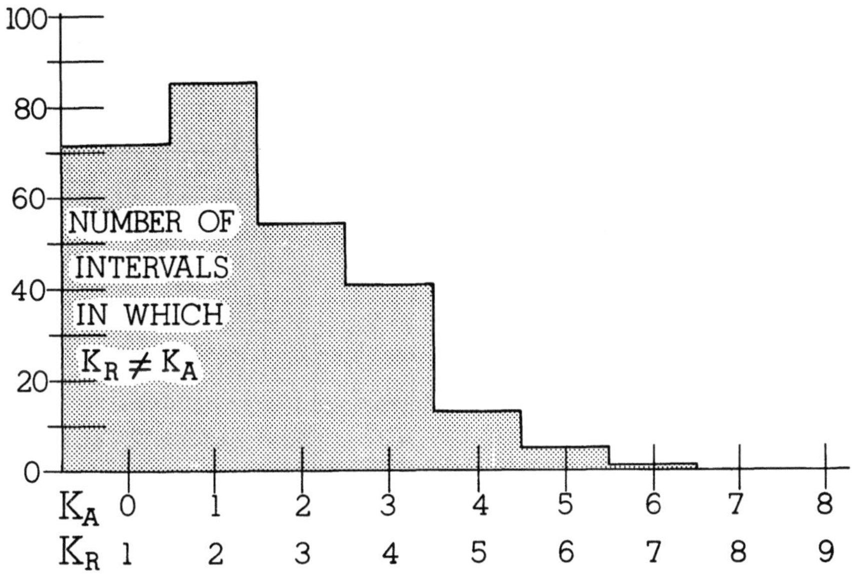

Fig. 5. Distribution of intervals of $K_A \neq K_R$ with the level of activity at Ottawa in 1982.

distribution (Figure 4) shows that unequal K values occur most frequently during the equinoxes and least frequently during the summer months. These distribution patterns are typical of those found at Ottawa for Pc4–Pc5 events (Gupta *et al.*, 1971; Gupta, 1973). Figure 5 shows the activity levels most affected by discrepant K values in this analysis. As would be expected these correspond to the lower K levels which contain much of the low energy pulsational activity with periods of 5 min or less.

3. Requirements for Satisfactory Derivation of *K* Index from Digital Data

Rapid geomagnetic fluctuations with periods less than about 200 sec are severely attenuated in AMOPLOTS constructed from 1 min averaged data. Small amplitude pulsations, sudden commencements, etc., which are readily identifiable on standard photographic magnetograms, are therefore either absent or much reduced on plots of this kind. This effect leads to a significant downward bias in K values measured from AMOPLOTS when activity levels are low (Figure 5).

In this study magnetograms were also reconstructed from the fast-sampled raw AMOS output for intervals of 60, 40, 30, 20, and 5 sec between successive data points (Figure 2). These fast-sample plots were produced for 90 three hour intervals between August 14 and December 31, 1982 in which discrepant K values were found between the Ruska magnetograms and the AMOPLOTS constructed from 1 min averaged data. However, in this case no averaging or filtering algorithms were applied to the digital data and K values were determined directly from the raw data plots. The results for this $4\frac{1}{2}$ month period are shown in Figure 6.

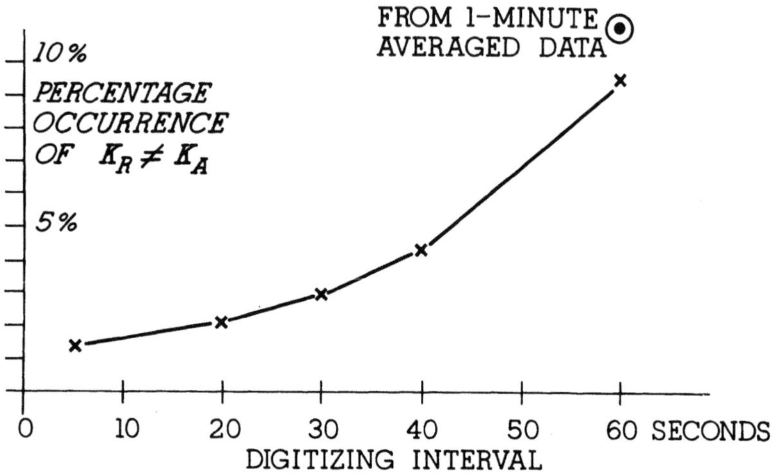

Fig. 6. Percentage occurrence of $K_A \neq K_R$ for 1 min averaged data (AMOPLOTS) and for plots constructed from spot values derived from the AMOS fast-sample data.

For the AMOPLOTS derived from 1 min averaged data it is again seen that measured K values are biased downward by one level in about 11% of the intervals. 80% of these low values occurred at the $K = 0, 1, 2, 3$, levels. The situation is only marginally better when 1 min spot values are used, but significant improvement is obtained when the magnetogram plots are constructed from digitizing intervals of 30 sec or less.

At these higher sampling rates the number of discrepant intervals is reduced to 3% or less. We estimate that an error level of this magnitude can be attributed either to observer errors (Menvielle, 1981) or to naturally occurring differences between K derived from H and D components and K derived from X and Y. The latter effect occurs much less frequently than the former.

We therefore conclude that digital data recorded at intervals of 30 sec or less are adequate for derivation of K indices at Ottawa. Such data are considered to have a reliability for K evaluation at least equivalent to that of standard Ruska analogue magnetograms. At longer digitizing intervals, however, inconsistencies are to be expected between K values derived from digital plots and those derived from traditional analogue recordings. One minute averaged values constructed along the lines recommended in IAGA Resolution 12 (Canberra, 1979) will clearly lead to discrepant K values at the lower disturbance levels.

References

Coles, R. L.: 1983, *Response functions for AMOS III Digital Filters*, Internal Report 83-2. Division of Seismology and Geomagnetism, Earth Physics Branch, Ottawa, Canada.

Gupta, J. C., Stening, R. J., and Jansen van Beek, G.: 1971, 'Micropulsations in the Pc3 to Pc4 Range at Four Canadian Observatories', *J. Geophys. Res.* **76**, 933–946.

Gupta, J. C.: 1973, 'Occurrence Studies of Geomagnetic Micropulsations Pc5 at High Latitudes', *J. Atmosph. Terres. Phys.* **35**, 2217–2235.
Mayaud, P. N.: 1965a, 'Analyse morphologique de la variabilité jour-a-jour de la variation journalière régulière S_R du champ magnetique terrestre. I. Le système de courants C_p (régions polaires et subpolaires)', *Annales de Geophysique* **21**, 369–401.
Mayaud, P. N.: 1965b, 'Analyse morphologique de la variabilité jour-a-jour de la variation journalière régulière S_R du champ magnetique terrestre. II. Le système de courants C_M (régions non-polaires)', *Annales de Geophysique* **21**, 514–544.
Menvielle: 1981, 'About the Scaling of K Indices', *IAGA News* No. 20, 110–111.
Trigg, D. F. and Nandi, A.: 1984, *The Automatic Magnetic Observatory System AMOS III*, Publications of the Earth Physics Branch, Geomagnetic Series, No. 27, Department of Energy, Mines and Resources, Ottawa, Canada.

THE GENERATION OF *K* INDICES FROM DIGITALLY RECORDED MAGNETIC DATA

J. C. RIDDICK and W. F. STUART
Geomagnetism Research Group, British Geological Survey, Edinburgh EH9 3LA

Abstract. The *K* index was developed by Bartels and is a measure of geomagnetic activity. The important feature of this index is that it is a measure of solar activity superimposed on the regular daily variation.

It is the most used estimate of activity being of particular interest to researchers in ionospheric physics and telecommunications.

In response to increased availability of digital data from magnetic observatories and their demand by the scientific community it becomes desireable to check whether *K* indices produced from magnetograms plotted from digital data are consistent with those handscaled from La Cour charts.

This paper presents results of these comparisons and also reports on several methods tested to produce the *K* index directly from the digital data.

Handscaled indices from digital plots are suitable for producing the *K* index. Computations using digital data do not reproduce *K* as defined by Bartels precisely but do provide an equivalent index which is satisfactory for most user and research purposes.

1. Introduction

The *K* index was introduced by Bartels *et al.* (1939) as an estimate of the level of geomagnetic disturbance due to corpuscular (as opposed to wave) radiation from the Sun. Bartel's way of separating the two fundamentally different sources of magnetic disturbance was to recognise that wave radiation results in the systematic external fields, notably the daily variation, and that the effects of corpuscular radiation are superimposed on an otherwise smooth daily variation. By subtracting the daily variation curve from a magnetogram the resulting level of activity of non systematic fields can be estimated. For a combination of reasons the index *K* is estimated as the range of the adjusted magnetic activity in three hourly intervals. Because geomagnetic activity has a very wide dynamic range at any given location a quasi-logarithmic scale was adopted to provide *K* values from 0 to 9 distributed in a way which was thought to be meaningful (Bartels *et al.*, 1939). Niemegk was the station at which the original description of the *K* index was developed and it has been used as the standard against which other observatories are calibrated. Because geomagnetic noise is a complex function of latitude (due to the variable dominance of discrete ionospheric/magnetospheric currents) the lower limit of $K = 9$ and the intermediate values are selected for each observatory so that the distribution of observed *K* values is consistent with that of *Niemegk*. Table I shows the levels of disturbance at Lerwick, Eskdalemuir, and Hartland corresponding to the values of *K* index. Mayaud (1980) examined *K* index scaling methods and describes uses of geomagnetic indices.

The *K* index is published monthly in several learned journals and is outstandingly the most used estimate of magnetic disturbance (in all fields of ionospheric and magnetospheric physics and in telecommunications and satellite navigation) in its

Geophysical Surveys **6** (1984) 439-456. 0046-5763/84/0064-0439$02.70.

TABLE I

Lerwick

Latitude: 60° 3′ N	Longitude: 358° 49′ E
Geomagnetic Latitude: +62.5	Geomagnetic Longitude: +88.6

Eskdalemuir

Latitude: 55° 19′ N	Longitude: 356° 48′ E
Geomagnetic Latitude: +58.5	Geomagnetic Longitude: +82.9

Hartland

Latitude: 50° 59.7′ N	Longitude: 355° 31′ E
Geomagnetic Latitude: +54.6	Geomagnetic Longitude: +79°.0

K Index	Lerwick	Eskdalemuir	Hartland
0	0–10	0–8	0–5
1	10–20	8–15	5–10
2	20–40	15–30	10–20
3	40–80	30–60	20–40
4	80–140	60–105	40–70
5	140–240	105–180	70–120
6	240–400	180–300	120–200
7	400–660	300–500	200–330
8	660–1000	500–750	330–500
9	1000+	750+	500+

original form or its global form Kp or ΣKp. Kp is derived from the mean of the individual standardized K values from 13 selected observatories around the world (the three U.K. observatories are included). Kp is presented on a scale of 27 rather than 9 by using a subscript notation for intermediate values, i.e. $0, 0_+, 1_-, 1_0, 1_+, 2_-, 2_0, 2_+$ etc. It has become popular to describe the level of magnetic activity for a given day as ΣKp, i.e. the sum of the 8 individual Kp values for the UT day. The ΣKp notation has often been extended to mean the 'sum of the Kp values for the hours under consideration'.

Because of its considerable value to the scientific community, and because there now exists a long data base of K scalings from most observatories (typically 40 yr, but K values have been scaled for Abinger back until 1929) it is very important to consider whether the evolution of recording practice at geomagnetic observatories will affect the continued production of consistent K estimations. This is a particular concern when digital variation recording is adopted. At the very least the characteristics of magnetograms plotted from digital records are different from those from classical analogue methods. The principal difference is the loss of high frequency response by the digital record but an important subjective element is the even trace thickness of digital plots at times of rapid field movement. Figure 1 shows a standard La Cour magnetogram and the corresponding computer plot for comparison.

In view of the arbitrary way in which the values of K are defined for each station and

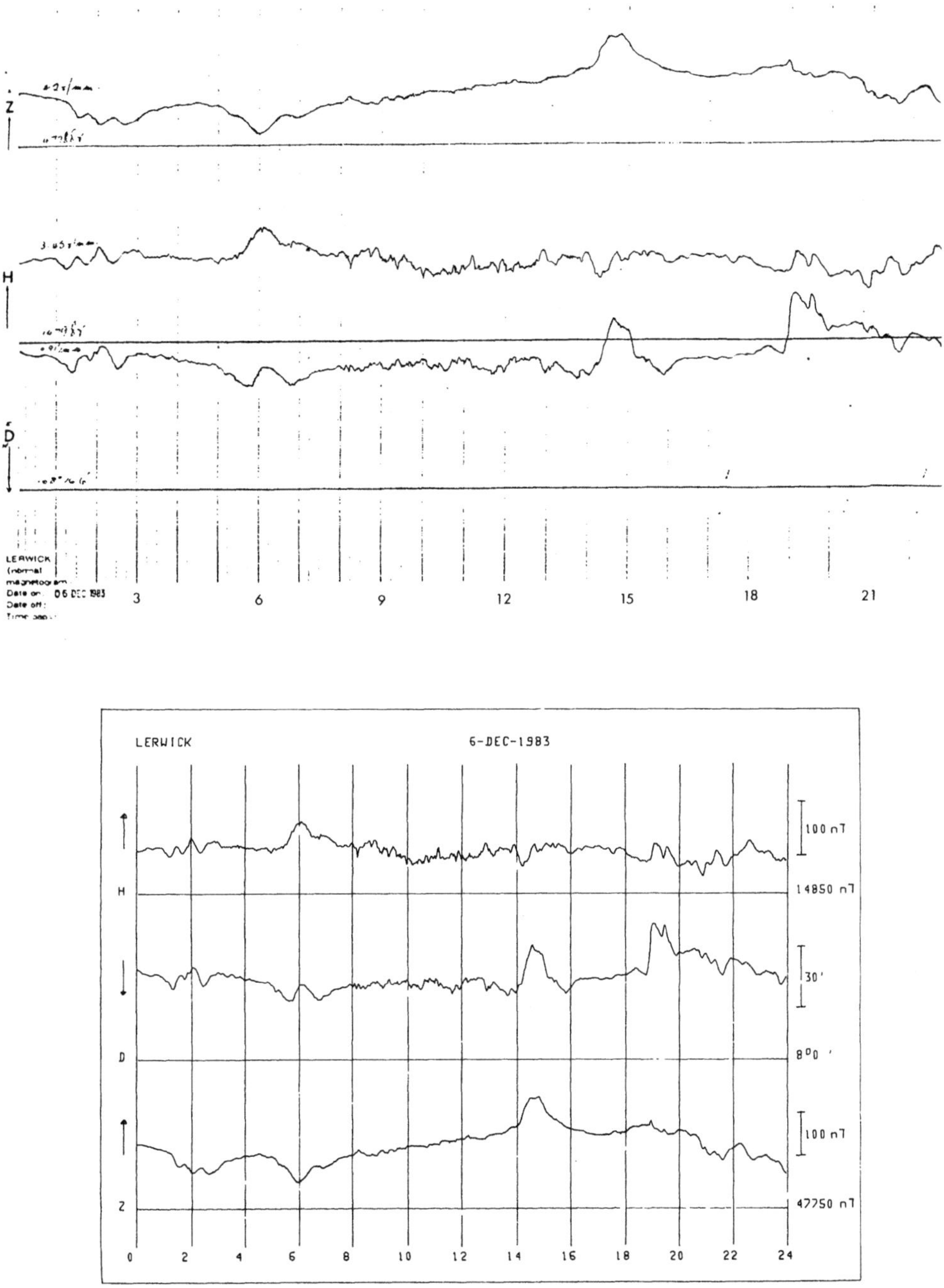

Fig. 1. A digital magnetogram and a La Cour photographic magnetogram for the same day from Lerwick Observatory.

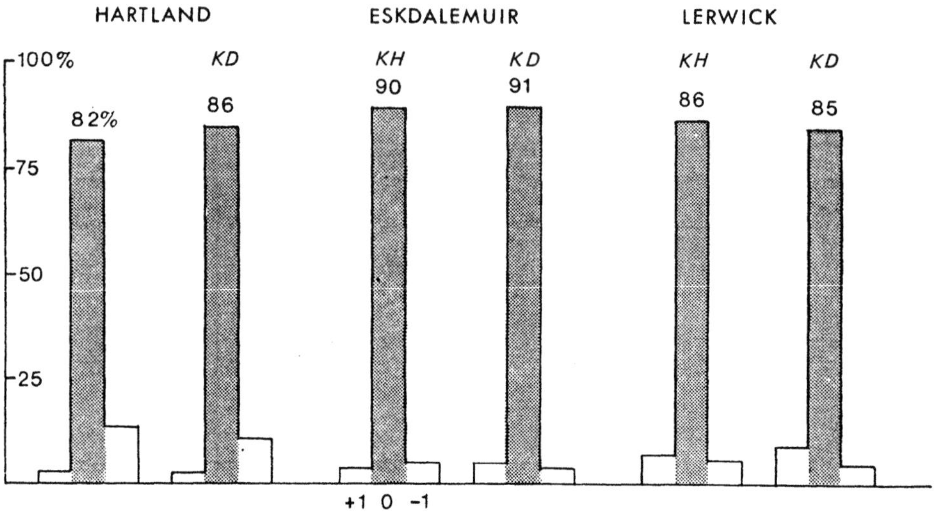

Fig. 2. Histogram showing the degree of agreement between the K scalings by two individual observers at each of the U.K. Observatories over a six month period. 0 is when both observers agree, + 1 is when observer A is 1 index value higher than observer B, − 1 is when observer B is 1 index value higher than observer A.

the advantages of operating directly with digital data, it is interesting to examine possibilities for computing 'quasi-K' indices directly and objectively from digital recordings. The problem is to derive an algorithm to subtract the daily variation in such a way that the computer generated K values are consistent with handscalings.

1.1. Handscaling

Since the UK observatories span a considerable range of latitudes it was thought necessary to include each station in the comparison (see Table I). At each observatory the standard La Cour magnetograms were handscaled independently by two experienced observers. The results are shown in Figure 2 and indicate that in about 85% of cases the two observers agree absolutely being only 1 K value different for all other cases. It is extremely rare for observers to differ by 2 increments of K value. This result confirms earlier work by Stuart *et al.* (1969) and should be compared with Loomer *et al.* (1983) who show 90% agreement in the situation where the two observers collude and adjust their estimates to agree in borderline cases. In the latter the 10% disagreement represents distinctly doubtful values within the overall 15% reproducibility of handscaled K estimates.

Handscaling of La Cour magnetograms are not examined further in this paper, only the main result is used as a standard by which to judge results from digital records.

1.2. Digital data

The digital recording and data processing system is described by Forbes and Riddick (1984) and Riddick *et al.* (1982). The magnetic field is sampled every 30 sec at a

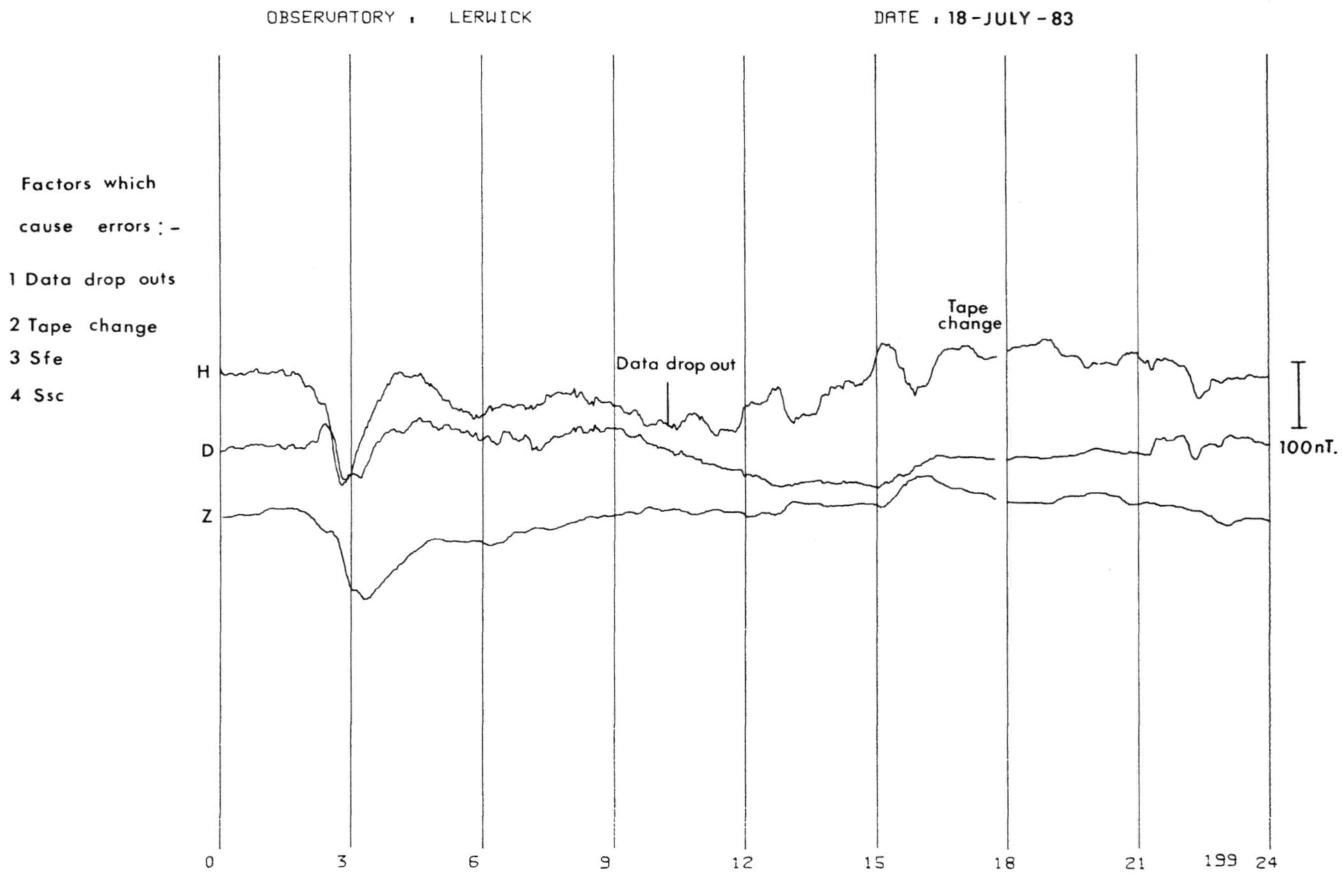

Fig. 3. A typical plot of a digital magnetogram to which has been added some possible sources of errors encountered using these plots. Different coloured ink is used for each component to overcome component identification problems when traces cross.

resolution of 1 nT. Magnetograms are plotted from minute 'mean values' obtained by averaging the value 'on the minute' together with the succeeding 30 sec sample. Minute mean values are computed to an accuracy of 1 nT. Different coloured inks are used for each component and pen thickness of the plots is 0.35 mm, the scale value of each component is set to be approximately 4 nT, similar to those of the observatory magnetograms. Figure 3 is an example of a plotted magnetogram illustrating some of the features which may contribute to errors. In Figure 3 the gap 17.55–18.00 UT is due to tape change which takes place once per week. A tape change itself takes only a matter of moments but results in the loss of 5 min of useful data due to the way data are blocked and the constraints in the processing procedure. Also shown is a data drop out at 10.20 UT. These can be caused by several things, e.g. mains transients, faulty tapes or dirty tape heads, and cannot be completely eliminated. The routine data process allows obvious ones to be identified on visual display of the magnetogram before plotting. It is usual to replace the incorrect data point by the mean of the data points on either side. See Mills *et al.* (1976) for more information about editing techniques. The resolution of 1 nT and 1 min samples combine with small rapid geomagnetic variations to produce a magnetogram which often has rectangular shapes instead of rounded. It is therefore not possible to distinguish data drop outs which only result in one or two nT errors from real and correct data. This is a source of error which cannot be eliminated without both higher time and field resolution.

When the cassettes data are transferred to the computer the first operation is to

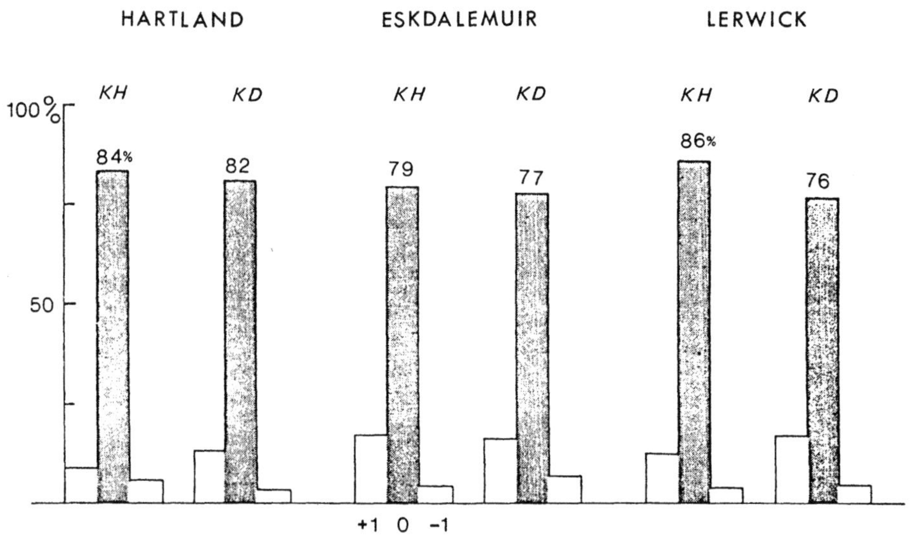

Fig. 4. Histogram showing a six month comparison of K, scaled from digital plots against K scaled from La Cour magnetograms at all three Observatories. 0 is when K derived from both charts agree, $+1$ is when K scaled from the La Cour chart is 1 value higher than K scaled from the digital plots, -1 is when the digital plot value is 1 higher than the La Cour.

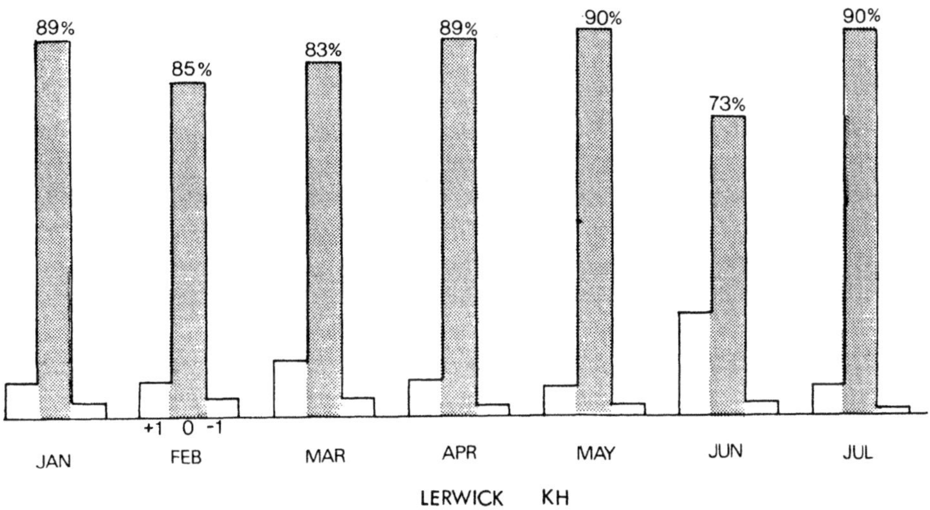

Fig. 5. Histograms showing the comparison of digital plots against La Cour derived indices over seven months at Lerwick, the +1, 0, −1 convention is as in Figure 4.

display the magnetogram on a VDU so that identifiable errors can be noted and if necessary edited. The magnetogram is then plotted on an on-line A–3 plotter. During 1983 the magnetograms were plotted on the same scale as those produced photographically by each observatory so that the normal *K* scale could be used for estimating *K* values from both types of magnetogram. At each observatory the observers estimated *K* values from the plots in the normal way. Figure 4 shows histograms of the comparison of the *K* values obtained for January-July, 1983 from the two types of magnetogram from all 3 observatories. Figure 5 shows seven individual monthly comparisons of indices derived from both methods at Lerwick and illustrate the range of variability between the two scalings. There is 99.5% agreement at the ± 1 level with 75–86% complete agreement. The two occasions when the difference exceeded 1 were traced to small artificial disturbances on the digital record in one case and a spectacular SSC (16.10 UT on 4th February 1983) which had a very large and rapid initial movement. The latter was only just detectable on the photographic record and was not faithfully recorded at the slow sampling of the digital record.

It is interesting to note that the distributions of 'errors' in Figure 4 are all biased slightly toward the +side, i.e. the photographic magnetograms produce greater *K* index than the plots. The bias is consistent at all observatories (cf. Figure 2, where Hartland showed a bias between observers) and is therefore thought to be real. Examination of the individual cases of difference show that they occur mainly in the range of *K*-values 0–2. This is probably because the slow sampling rate of the present plotted magnetograms (1 min) does not record rapid variations faithfully. The result is consistent with that of Loomer *et al.* (1983) who demonstrate the frequency response of sampled data and suggest that a 30 sec sampling rate has an effective amplitude response of 95%. Loomer *et al.* find a similar bias toward higher K values from photographic records and conclude that the data sampled at 30 sec excludes Pc4–5 activity at the latitude of the Canadian stations.

The result of this test is very encouraging. K index is a classification rather than a precise measurement and the comparisons show that plotted digital magnetograms are satisfactory for the production of indices which are consistent, within the range of normal discrepancy, with those prepared from traditional recording methods. There is no systematic difference between handscaled K values from plots and magnetographs apart from the slight tendency of high values from magnetograms. Plotted magnetograms can be produced within 72 hr of the cassette change at the observatories and there will be no reduction in quality of the service to the user community. Some users require an indication of geomagnetic activity 'as soon as possible'. Enquiries have shown that these requirements are satisfied by general level of activity index, typically to answer the questions, is $K > 3$, $3 < K < 6$, $K > 6$, (i.e. more like the 'C' index which was a daily magnetic character 1, 2 or 3). These requirements can be met by the use of a low quality on line pen recorder.

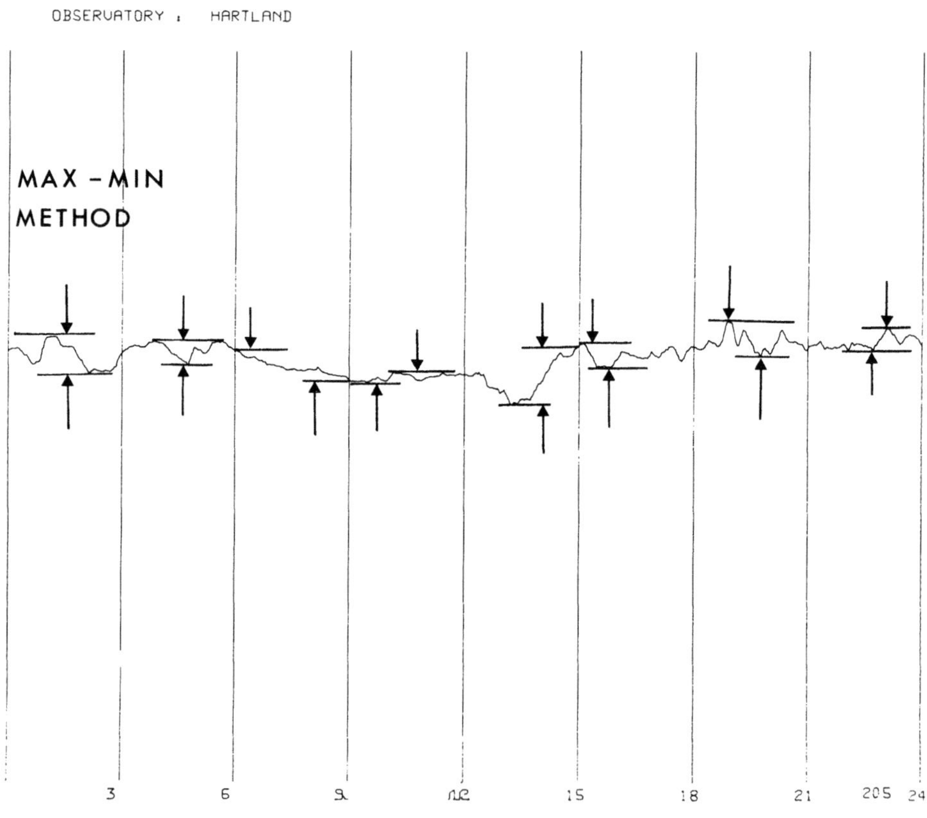

Fig. 6. A magnetogram demonstrating how indices are obtained using the MAX-MIN Method.

1.3. Automatically produced quasi-K index

The main element in estimating K is elimination of the smooth daily variation (S_R). The recommended method for doing this is for an observer experienced in the local form and variation of the daily curve to use his judgement. The daily curve varies systematically with latitude, season and level of magnetic activity (see Chapman and Bartels, 1940). There are also non-systematic variations due to minor changes in the topology of the current system in the daylight hemisphere which causes it. These are most important at the latitude of the 'focus' where small variations in focus latitude cause large changes in the form of the magnetic variation. At the latitude of the U.K. observatories small changes in the topology of the current system have only very small effects on the daily magnetic variation and it is to be expected that only the systematic seasonal variations need to be taken into account.

An attempt was, therefore, made to synthesise an algorithm which could be used for automatic production of the K index. The first step was to test the simple three hourly range or magnetic activity without consideration of daily variation.

1.4. The max-min method

Figure 6 illustrated the method of estimating the three hourly range of magnetic activity. The digital data are scanned for maximum and minimum in each three hour interval and the resulting range in nT is converted to K values using the conversion codes shown in Table I. Indices can be produced in this way very quickly with almost no cost in computing time.

Comparisons of Max-Min quasi-K indices against handscaled indices from La Cour magnetograms are shown for the UK observatories in Figure 7a. Lerwick data cover a six month period, data from Eskdalemuir and Hartland are presented for one month along with the corresponding month at Lerwick for comparison (Figure 7b). The monthly distributions at each observatory are similar with Lerwick showing a 68–75% agreement, Eskdalemuir 80% and Hartland 60%. At Lerwick and Eskdalemuir the agreement at the $+1$ level is better than 99% at Hartland 95%. The most striking feature of the comparison is the tendency for the method to produce high indices particularly at Hartland with 34% of the digitally produced indices 1 value higher than the corresponding La Cour index. At Lerwick and Eskdalemuir this bias to higher digital values is reduced to 15%.

Since the automatic index is computed objectively without regard to station sensitivity and range of $K = 9$ the variation between observatories may reflect real physical differences in the level or morphology of K 'noise' at the three stations, bias created by the combination of classical magnetogram scale value and the magnitude of the increments of K value or bias of the local observers. It is not the purpose of this paper to investigate these possibilities in detail but it is worth remarking that at Hartland the bottom line for $K = 9$ is 500 nT and the scale value of magnetograms is 4.34 nT mm^{-1} while at Lerwick $K = 9$ is 1000 nT and the scale value is 3.45 nT mm^{-1}. Trace thickness at Hartland is 0.5 mm, corresponding to 2.2 nT. At Lerwick it is

0.4 mm, corresponding to 1.4 nT. This combination of error limits in the basic traces suggest the expectation that, particularly for low K values, the distribution of K value discrepancies will be broader at Hartland than Lerwick. Figure 7b suggests that this is the case, but, surprisingly, the classical versus plotted comparison of Figure 4 does not show the same effect.

In principle, the fact that S_R is not taken into account in the Max-Min method should result in consistently high K values being estimated because the background slope is not taken into account. Figure 8 shows the form of the background S_q field and it is clear that between 18.00 and 06.00 UT there is very little curvature or slope in the background field to be removed. The main slope (and its variability occurs between 06.00 and 18.00, predicting that night time Max-Min results should be consistently better than daytime values. This is illustrated in Figure 11 where the daily variation of

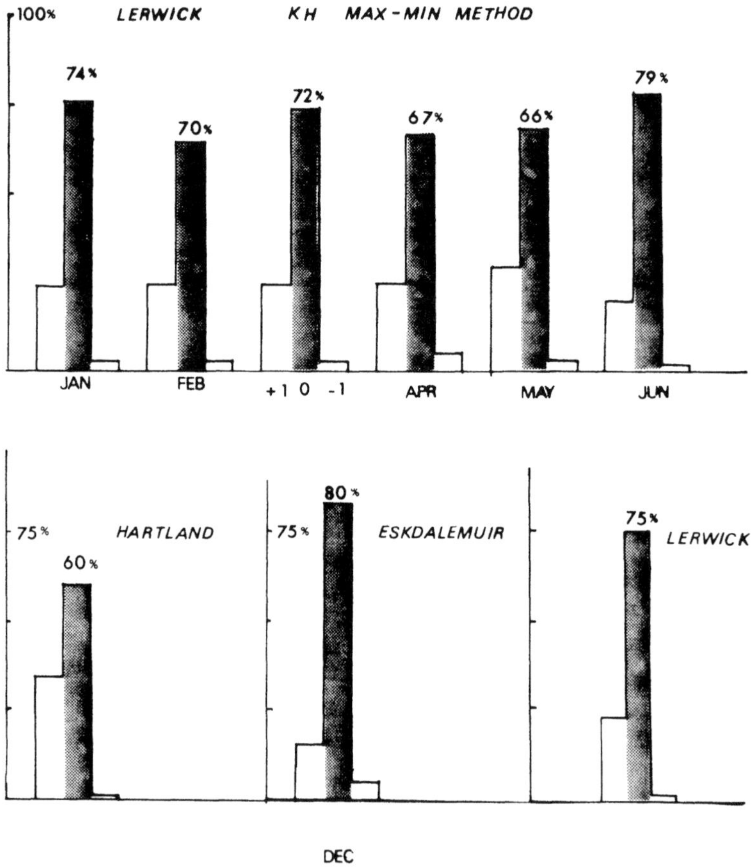

Fig. 7. Histograms from all observatories giving the results of the comparisons of the indices derived using MAX-MIN method against the La Cours. 0 is when both indices agree, −1 is when the La Cour index is 1 value higher than the digital index. +1 is when the digital index is 1 value higher than the La Cour.

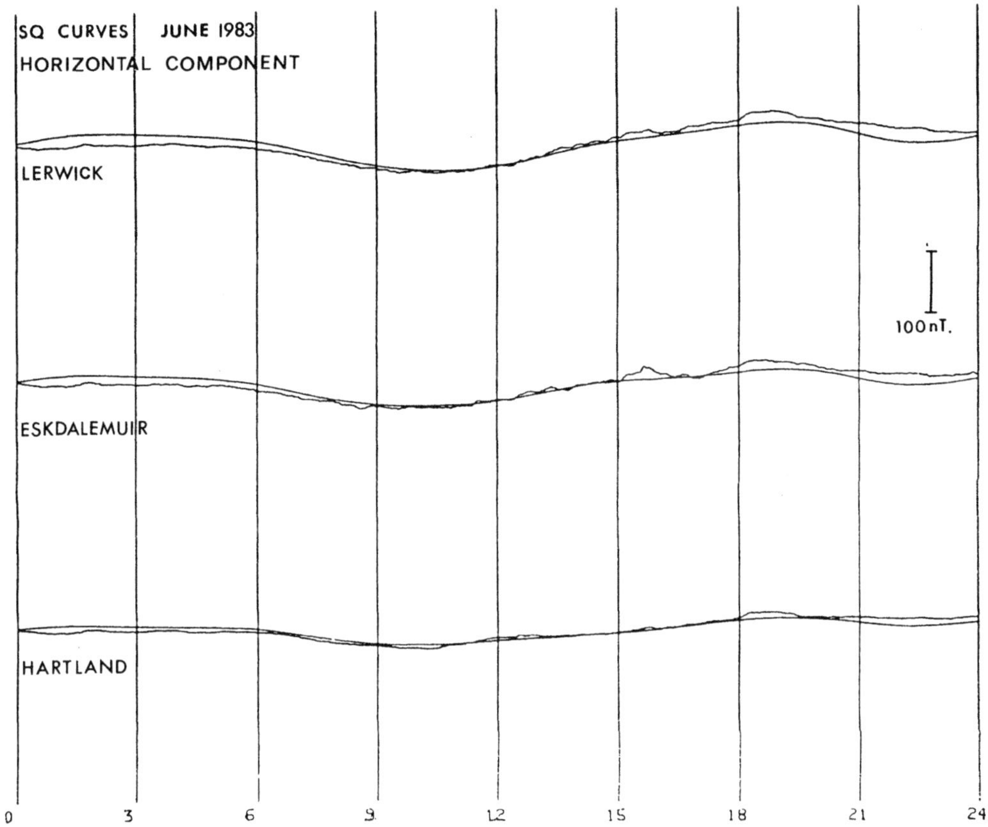

Fig. 8. S_q curves for the horizontal component at all three observatories derived from five magnetically quiet days in June 1983 using the mean S_q and Harmonic Synthesis (smooth curve) methods.

'agreement' of the Max-Min method is shown together with that of the other automatic methods. It is clear from Figure 11 that most of the discrepancies occur between 06.00 and 18.00 UT.

In an attempt to eliminate these errors each 3-hr interval was linearly detrended and the slope of the resulting straight line was used as a reference from which the range was then estimated. A regression line of the form $y = a_1x + a_0$ was fitted to the 30 sec data for each 3-hr interval by the method of least squares and then subtracted from the original data to find the maximum and minimum deviation from it. This adjusted range was then converted to K values as before. The method resulted in poorer agreement with handscalings (approximately 50% complete agreement). Examination of the recordings used brought out the fact that the least squares straight line is a very inadequate representation of the geomagnetic field over a three hour period, primarily because disturbance amplitude increases rapidly with its time scale so that activity with time scales of 30 min or more can and do bias the regression line away from the quiet field value.

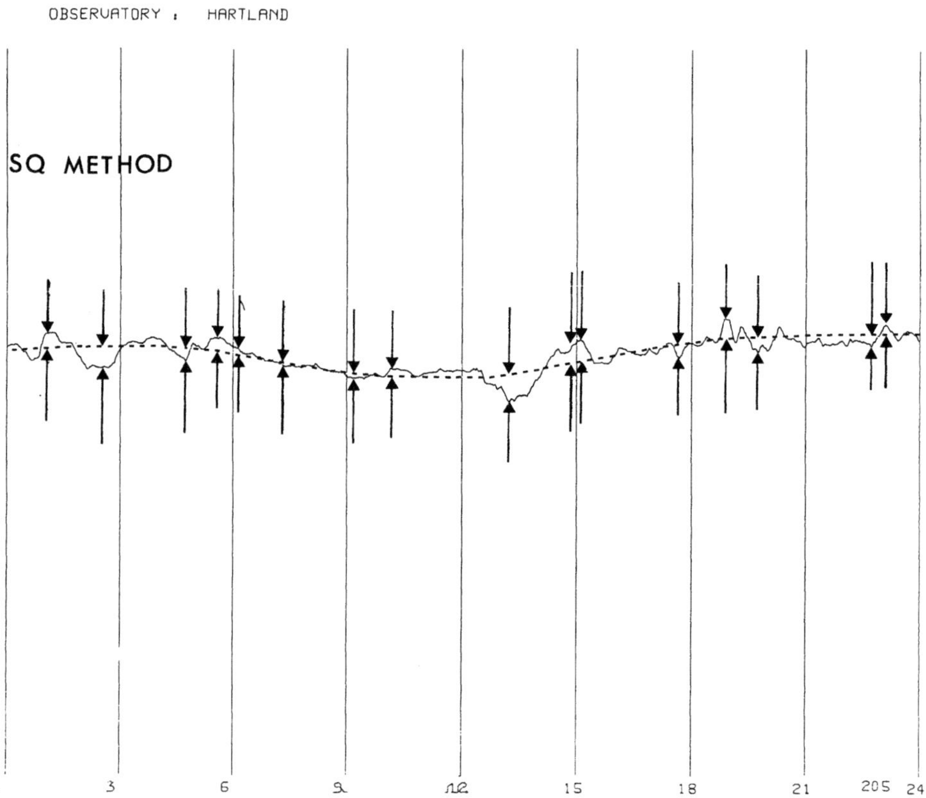

Fig. 9. A digital magnetogram showing how the SQ and SQF methods of deriving the *K* indices are applied.

1.5. Mean daily variation curve

A standard procedure of magnetic reporting is to select the 'five quiet' days of each calendar month. These are defined as the quietest days in the month so are relatively rather than absolutely quiet. In order to take some account of daily variation in automatic *K* estimation and to eliminate the diurnal error effects of the 3-hr range (Max-Min) method, a set of curves was computed by averaging the five quiet days (i.e. the five sets of 30 sec data for the selected five quiet days were meaned measurement by measurement) to produce a mean quiet day variation curve for the month. The meaned data were then subtracted from the 30 sec data for each day to produce an adjusted range and *K* indices evaluated as before. The method of estimating *K* values from the daily variation curve is shown in Figure 9. This method will be referred to as the *SQ* method.

A second method of obtaining an S_q curve was then tried. Again the same five quiet days for the month were selected and a curve obtained from synthesis of the coefficients of the first five harmonics of the daily spectrum. The curves produced by this method (SQF) and those produced by the *SQ* method are shown in Figure 8. As it is not possible

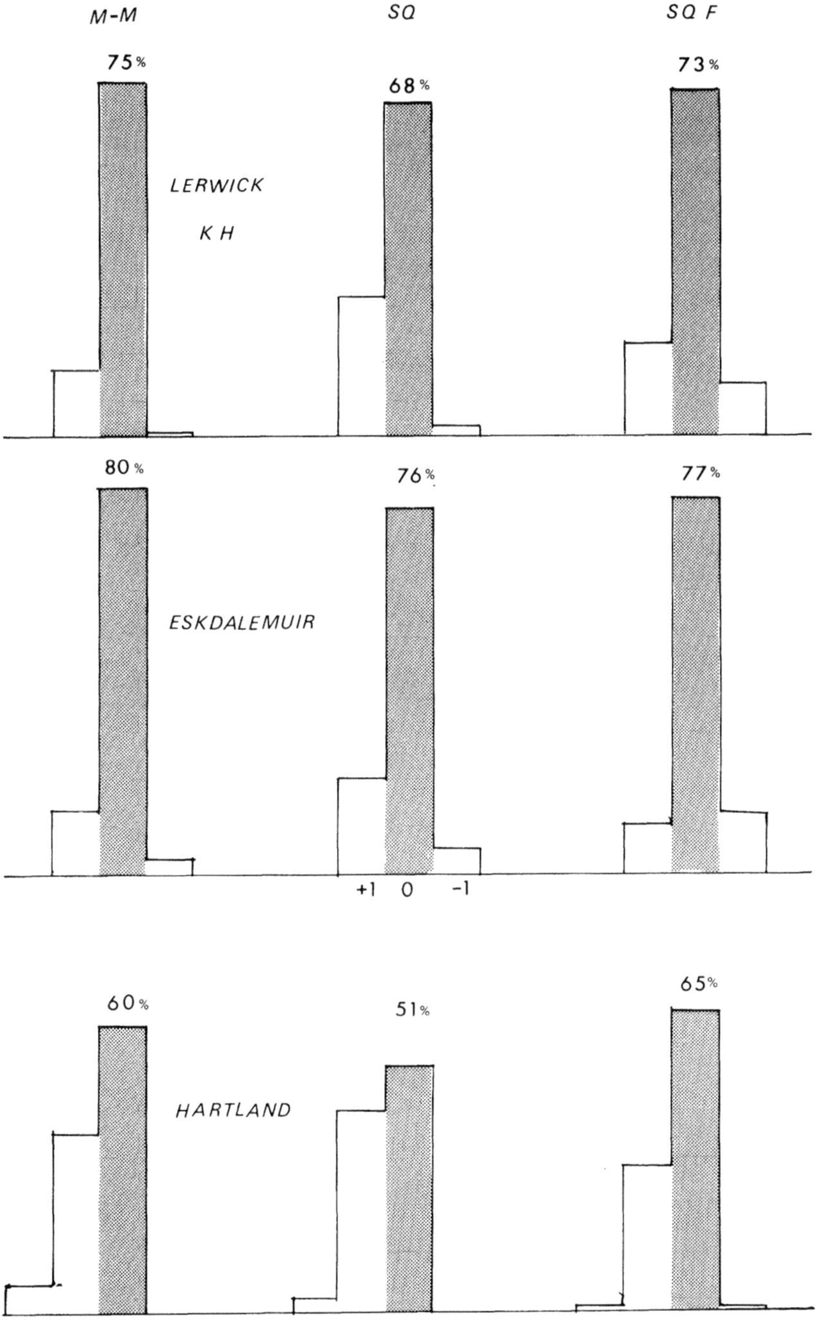

Fig. 10. Histograms giving the results of comparisons between digitally derived indices using Max-Mln, SQ, and SQF methods and the indices derived from La Cour magnetograms. 0 is when all methods agree, + 1 is when the digital K is 1 index value higher than the La Cour, − 1 is when the La Cour K is one value higher than the digital K.

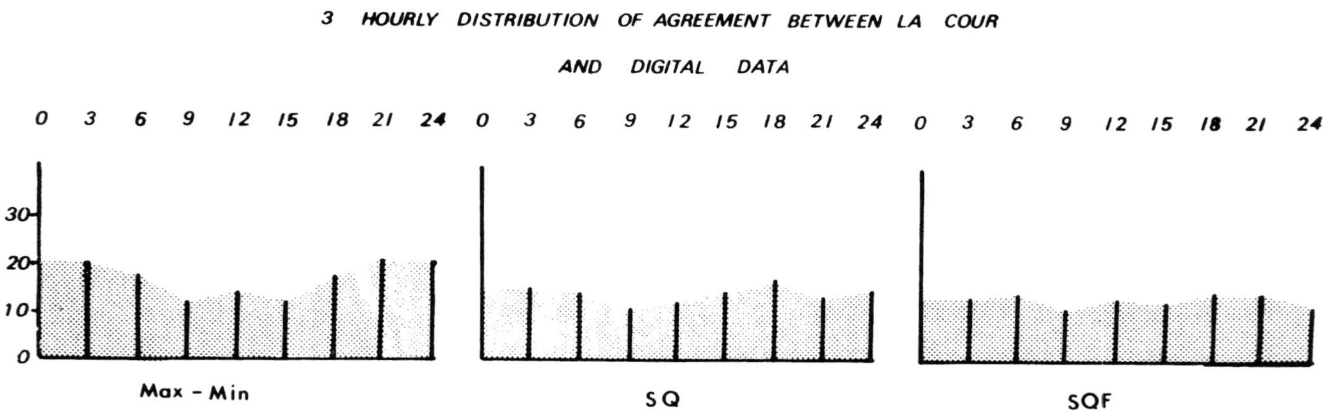

Fig. 11. Histograms showing the distribution of agreements in relation to the three hour K interval. The results of K derived by all three digital methods is compared to the K scaled from the La Cours. The vertical scale gives the number of agreements between the La Cour indices and the Digital indices.

to obtain a daily variation curve for each day (i.e. S_R) the curves used by the digital methods are representative curves for each month which in the past have been referred to as 'iron curves'.

Again this curve the was subtracted from 30 sec data and the K indices derived as with the other methods. Figure 10 shows a summary of the comparison of the three digital methods with handscalings from classical magnetograms the data for Lerwick covers a six month period, Eskdalemuir and Hartland data is one representative month for the horizontal component. The three methods produce very similar overall statistics. Figure 11 shows the daily range of errors. The Max-Min method has a low level of agreement between 06.00 and 18.00 UT.

The curves shown in Figure 8 illustrate the main shortcomings of using selected quiet days without regard to the absolute level of noise. In the meaned curves (SQ) the effects of activity on one day is clearly shown between 15.00 and 21.00 UT where the traces have quite large digressions. Since these traces are the basis for estimating disturbance it is clear that for the month in question all of the ranges from which indices are produced for 15.00-18.00 UT and 18.00-21.00 UT will be in error. For 15.00-18.00 the error at Lerwick is about 10 nT, at Eskdalemuir it is about 20 nT and at Hartland 3–4 nT. Errors due to this cause affect K estimates at all levels if the estimate is close to the increment value and may well result in badly wrong estimates for $K = 0$, 1 or 2.

By comparison the synthesised curve is smooth and a fairly good representation of the mean curve except between 00.00 to 03.00 UT and 21.00 to 24.00 where the harmonic analysis produces artificial curvature at the end of the record. In Figure 11 the SQF results shows this reduction in agreement at 0 and 24 UT.

2. Discussion

The results from the comparison of estimates for handscaled digital plots versus the La Cour handscalings were consistent with the results obtained when independent experienced observers scaled La Cour magnetograms and 75–90% agreement was achieved at all stations with 99.5% agreement between ± 1. The most significant error appears to be a tendency for the La Cour indices to be higher than those produced from the digital plots. This can be corrected by increasing the digital sampling rate (Loomer *et al.*, 1983). It was also noted that there was a marked reduction in the overall percentage agreement in June 1983 at Lerwick (Figure 5) where there was an exchange of observers between Lerwick and Eskdalemuir. In previous tests (Figure 2) the observers at Eskdalemuir show a high degree of agreement as did the observers at Lerwick. Possibly observers become biassed to the particular characteristics of the range of geomagnetic activity and the S_R curve at each observatory.

The indices obtained by digital Max-Min method were in reasonable agreement with those derived from La Cour scalings. At Lerwick over the six month period there was a 72% average agreement. The 80% agreement obtained on the month tested at Eskdalemuir was very encouraging and there is no reason to expect it to vary to any great extent (the range of the Lerwick monthly comparisons using the Max-Min

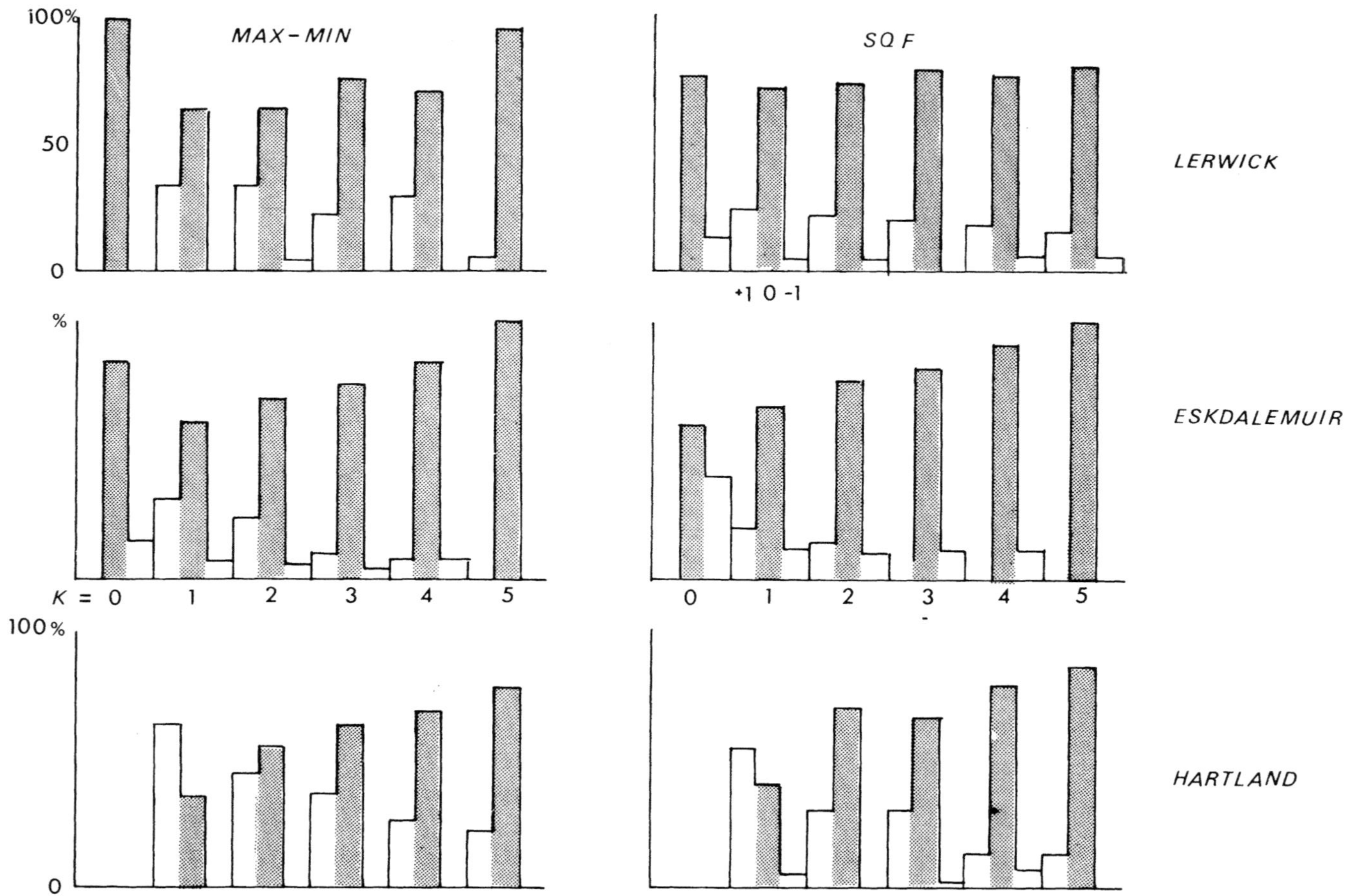

Fig. 12. Histograms giving the percentage agreements in the ranges $K = 0$ to 5 between Digital and La Cour derived indices for Max-Min and SQF methods using the digital data as a reference. + 1 is when the digitally derived index is 1 value higher than the La Cour, the shaded area is when the La Cour and digital index agree.

method is shown in Figure 7a to be 6%). Disagreement in low indices, $K = 0$ or 1 at Hartland, which produced only a 60% agreement, were the main reason for this slightly lower level of agreement at this observatory. These higher digital indices may be the result of electrical noise transients 1 or 2 nT in amplitude superimposed on the magnetic variations where the $K = 0$ range is only 5 nT. A fault in the mains wiring at this observatory has been detected which would account for an increase in artificial site noise.

Figure 12 shows the percentage agreement for indices in the range $K = 0$ to $K = 5$ for six months at Lerwick and one month at Eskdalemuir and Hartland. The shaded areas represent the degree of agreement between digital and handscaled indices, with the disagreement decreasing with increasing K value to be within reasonable limits at $K = 4$ or 5. It can be seen from these histograms that there are no indices at Hartland in the range $K = 0$. To overcome this tendency for digital analysis of low latitude data to produce higher indices at low K values, it may be necessary to decrease not only the sampling interval to 10 sec but also the recording resolution to 0.5 nT or better.

The S_q methods appear to reduce the systematic error of the simple Max-Min method but the improvement is only a few percent overall. The uncorrected Max-Min method for the automatic generation of K-indices would, however, appear to be a satisfactory method for the production of K indices for users requiring a general indication of activity. Although the K index is a value estimated subjectively by experienced observers with a specialist knowledge of local effects and S_q variations, it does seem possible to derive an algorithm to accommodate these effects for all latitudes and for the full range of S_q variations. The methods shown are probably acceptable for most geomagnetic purposes. Possibly with the increased number of observatories producing digital data an alternative 'quasi K' index should be allocated, this index being derived from a standard software package, the computational procedures of which could be agreed by an international panel in the same way as the International Geomagnetic Reference Field (IGRF) is defined.

3. Conclusion

Magnetograms derived from digital data sampled at 30 sec intervals will be used for the production of K indices from the U.K. observatories from January 1984.

Indices produced directly from digital data by computational processes using the techniques tested, exhibit a high enough overall degree of agreement with La Cour derived indices to be acceptable for many purposes.

Acknowledgments

The authors wish to express their gratitude to the staff of the UK Geomagnetic Observatories who provides much of the data used in the preparation of this report. This report is published by permission of the Director, British Geological Survey (N.E.R.C.).

References

Bartels, J., Heck, N. H., and Johnston, H. F.: 1939, 'Three Hour-range Index Measuring Geomagnetic Activity', *Terr. Magn. Atmos. Electricity.*

Chapman, S. and Bartels, J.: 1940, *Geomagnetism*, Oxford University Press.

Forbes, A. J. and Riddick, J. C.: 1984, 'The Digital Recording System operated at the U.K. Observatories', *Geophysical Surveys* **6**, 393–405 (this issue).

Loomer, E. I., Janson Van Beek, G., and Niblett, E. R.: 1983, *Derivations of K-indices Using Magnetograms Constructed from Digital Data*, *IAGA News* No. 22.

Mayaud, P. N.: 1980, 'Derivation, Meaning and Use of Geomagnetic Indices', *Geophysical Monograph* **22**, American Geophysical Union.

Mills, P. M., Dyson, K., Green, C. A., and Stuart, W. F.: 1976, The Data Transcription and Retrieval System used for Cassette Recording of the I.G.S. Rubidium Magnetometers, *Geomagnetism Unit Report* No. 20.

Riddick, J. C., Forbes, A. J., and Green, C. A.: 1981, 'The Recording and Processing of Digital Magnetic Data from the U.K. Observatories', *Geomagnetism Unit Report* No. 27.

Stuart, W. F., Dyson, K., and Mackintosh, S. M.: 1969, 'The Use of Semi-automatic Scaling Methods in Observatory Geomagnetism', *Geomagnetism Unit Report* No. 4.

ANNOUNCEMENT

Geomagnetic Observatory and Survey Practice

Editor: W. F. Stuart

Please note that a hardbound edition of these special issues of *Geophysical Surveys*, Vol. 6 Nos. 3 and 4 (July/October 1984), is available from the publishers.

ISBN 90–277–1908–X. Prices: Dfl. 105,– / $39.00 / £26.75.

TABLE OF CONTENTS

ARTICLES

ALABI, A. O. / Magnetometer Array Studies 153

AUSTER, V. / Geomagnetic Absolute Measurements with a Nuclear Resonance Theodolite at the Adolf Schmidt Observatory in Niemegk 269

AUSTER, V. and K. LENGNING / Comparison of Standard and Digital Observation at the Geomagnetic Adolf Schmidt Observatory Niemegk 351

BEBLO, MARTIN, *see* Schulz, Günter

BERGMARK, TORSTEN / Experience of Geomagnetic Field Recording with a Fluxgate Magnetometer Having a Bridge Sensor 381

BERKTOLD, A. / Electromagnetic Studies in Geothermal Regions 173

BITTERLY, J., J. M. CANTIN, R. SCHLICH, J. FOLQUES, and D. GILBERT / Portable Magnetometer Theodolite with Fluxgate Sensor for Earth's Magnetic Field Component Measurements 233

CANTIN, J. M., *see* Bitterly, J. *et al.*

CLERC, G., J-P. DÉCRIAUD, G. DOYEN, M. HALBWACHS, M. HENROTTE, J. RÉMY, and X-C. ZHANG / An Automatic Audio-Magnetotelluric Equipment, Controlled by Microprocessor, for the Telesurveillance of the Volcano Momotombo (Nicaragua) 291

COLES, R. L., *see* Niblett, E. R. *et al.*

DÉCRIAUD, J-P., *see* Clerc, G. *et al.*

DOSSO, H. W. and J. T. WEAVER / Preface 1

DOYEN, G., *see* Clerc, G. *et al.*

DRIMUSZ, L., *see* Hegymegi, L.

FOLQUES, J., *see* Bitterly *et al.*

FORBES, A. J. and J. C. RIDDICK / The Digital Recording System Operated at the U.K. Magnetic Observatories 393

FRASER, B. J. and P. W. McNABB / The Recording of Pc1 Geomagnetic Pulsations Using a Microcomputer Preprocessing System 249

GILBERT, D., *see* Bitterly, J. *et al.*

HALBWACHS, M., *see* Clerc, G. *et al.*

HEGYMEGI, L. and L. DRIMUSZ / An Intelligent Digital Magnetic Recording System (Dimars) 317

HENROTTE, M., *see* Clerc, G. *et al.*

HOHMANN, GERALD W. / Three-Dimensional EM Modeling 27

JANKOWSKI, J., J. MARIANIUK, A. RUTA, C. SUCKSDORFF, and M. KIVINEN / Long-Term Stability of a Torque-Balance Variometer with Photoelectric Converters in Observatory Practice 367

JANSEN VAN BEEK, G., *see* Niblett, E. R. *et al.*

JONES, ALAN G. / The Problem of Current Channelling: A Critical Review 79

JOSELYN, J. A. / Proposed Major Format Change to Geomagnetic Activity Reports and Forecasts Produced by the SESC, Boulder, Colorado, U.S.A. 419
KANEKO, T., *see* Utashiro, S. *et al.*
KIVINEN, M., *see* Jankowski, J. *et al.*
KLÖCKER, N., *see* Lühr, H. *et al.*
KOVAC, I., *see* Podsklan, J.
KUWASHIMA, M. and Y. SANO / Improved Kakioka Automatic Standard Magnetometer (KASMMER) 357
LAŠTOVIČKOVÁ, M. / Laboratory Measurements of Electrical Properties of Rocks and Minerals 201
LAW, L. K. / Marine Electromagnetic Research 123
LAURIDSEN, E. KRING / Absolute Measurement of *D* by Means of a Proton Magnetometer 223
LENGNING, K., *see* Auster, V.
LENNERS, D., *see* Linthe, H.-J.
LINTHE, H.-J. and D. LENNERS / Acquisition and Primary Processing of Geomagnetic Pulsations in Real Time Operation by Microcomputer 241
LOOMER, E. I., *see* Niblett, E. R. *et al.*
LÜHR, H., S. THÜREY, and N. KLÖCKER / The Eiscat-Magnetometer Cross 305
LUNDBAK, ASGER / About Digital Alternative to the *Kp*-Index 425
MARIANIUK, J., *see* Jankowski, J. *et al.*
McNABB, P. W., *see* Fraser, B. J.
MELONI, A., F. MOLINA, P. PALANGIO, Q. TACCETTI, and ANNA DE SANTIS / Automatic Digital Recording of Geomagnetic Elements by Means of a Proton Precession Magnetometer 339
MOLINA, F., *see* Meloni, A. *et al.*
NIBLETT, E. R., E. I. LOOMER, R. L. COLES, and G. JANSEN VAN BEEK / Derivation of *K*-Indices Using Magnetograms Constructed from Digital Data 431
OSHIMA, S., *see* Utashiro, S. *et al.*
PALANGIO, P., *see* Meloni, A. *et al.*
PARKER, ROBERT L. / The Magnetotelluric Inverse Problem 5
PODSKLAN, J. / Magnetic Mapping of Slovakia for the Epoch 1980.5 261
PODSKLAN, J. and I. KOVAC / One Year Experience with the CMVS 2 Magnetic Variation Station 335
RÉMY, J., *see* Clerc, G. *et al.*
RIDDICK, J. C. and W. F. STUART / The Generation of *K* Indices from Digitally Recorded Magnetic Data 439
RIDDICK, J. C., *see* Forbes, A. J.
ROUSE, J. R., *see* Simmons, D. A.
RUTA, A., *see* Jankowski, J. *et al.*
SANO, Y., *see* Kuwashima, M.
SANTIS, ANNA DE, *see* Meloni, A. *et al.*
SCHLICH, R., *see* Bitterly, J. *et al.*
SCHULZ, GUNTER and MARTIN BEBLO / On the Reduction of Time Variations for Geomagnetic Repeat Station Measurements 323

TABLE OF CONTENTS

SIMMONS, D. A. and J. R. ROUSE / Geomagnetic Measurements Made on the Moving Ice Shelf at Halley, Antarctica 407
STUART, W. F. / Introduction 217
STUART, W. F., *see* Riddick, J. C.
SUCKSDORFF, C., *see* Jankowski, J. *et al.*
TACCETTI, Q., *see* Meloni, A. *et al.*
THÜREY, S., *see* Lühr, H. *et al.*
UTASHIRO, S., S. OSHIMA, and T. KANEKO / Aeromagnetic Surveys in the Seas Around Japan in 1980 271
VARENTSOV, Iv. M. / Modern Trends in the Solution of Forward and Inverse 3D Electromagnetic Induction Problems 55
WARD, STANLEY H. / Controlled Source Electrical Methods for Deep Exploration 137
WEAVER, J. T., *see* Dosso, H. W.
ZHANG, X-C., *see* Clerc, G. *et al.*

Announcement 216, 457

Erratum 215

Sediment Diagenesis

Proceedings of the NATO Advanced Study Institute on Sediment Diagenesis, Reading, U.K., July 12–25, 1981

Edited by
A. PARKER and B. W. SELLWOOD

1983, vii + 427 pp.
Cloth Dfl. 140,– / US $ 53.50 ISBN 90-277-1677-3
NATO ADVANCED SCIENCE INSTITUTES SERIES
C. Mathematical and Physical Sciences 115

The topic of sediment diagenesis is of fundamental importance to industry in the evaluation of hydrocarbon and water reservoir rocks. Detailed knowledge of the diagenetic textures, fabrics, and minerals, and a prediction of the regional diagenetic response, partly controls hydrocarbon recovery programmes. In other words, knowledge of the diagenesis can aid (or even control) conservation policy. Similarly, facies and diagenetic trends within basins can influence exploration policy. In this volume, the major themes of carbonate and terrigenous clastic sediments were treated sequentially from deposition to deep burial. The keynote addresses which acted as the touchstones for discussion are presented here in the expectation that they will stimulate a still wider audience.

Contents: Foreword. T. Elliott: Facies, Sequences and Sand-Bodies of the Principal Clastic Depositional Environments. G. V. Chilingarian: Compactional Diagenesis. K. Bjørlykke: Diagenetic Reactions in Sandstones. B. Velde: Diagenetic Reactions in Clays. H. Fuchtbauer: Facies Controls on Sandstone Diagenesis. N. P. James: Depositional Models of Carbonate Rocks. R. G. C. Bathurst: Early Diagenesis of Carbonate Sediments. H. R. Wanless: Burial Diagenesis in Limestones. Index

D. Reidel Publishing Company

P.O. Box 17, 3300 AA Dordrecht, the Netherlands
190 Old Derby St., Hingham, MA 02043, U.S.A.

GPSR Compliance

The European Union's (EU) General Product Safety Regulation (GPSR) is a set of rules that requires consumer products to be safe and our obligations to ensure this.

If you have any concerns about our products, you can contact us on ProductSafety@springernature.com

In case Publisher is established outside the EU, the EU authorized representative is:

Springer Nature Customer Service Center GmbH
Europaplatz 3
69115 Heidelberg, Germany

Batch number: 10607222

Printed by Printforce, the Netherlands